# The Role of Salicylic Acid and Nitric Oxide in Plant Heat Response

# The Role of Salicylic Acid and Nitric Oxide in Plant Heat Response

By

Krishna Kumar Rai

**Cambridge Scholars**
Publishing

The Role of Salicylic Acid and Nitric Oxide in Plant Heat Response

By Krishna Kumar Rai

This book first published 2023

Cambridge Scholars Publishing

Lady Stephenson Library, Newcastle upon Tyne, NE6 2PA, UK

British Library Cataloguing in Publication Data
A catalogue record for this book is available from the British Library



ISBN (10): 1-5275-4073-1
ISBN (13): 978-1-5275-4073-6

# TABLE OF CONTENTS

# CHAPTER 1

# HIGH-TEMPERATURE STRESS AND ITS IMPACT ON AGRICULTURE - CURRENT SCENARIO


## Abstract

High temperature (HT) stress is one of the most devastating environmental stressors that affect agricultural productivity and global food security. HT stress imposes severe repercussions on crop plants and livestock growth and productivity. HT stress provokes stigma exertion and drastically reduces pollen efficiency, leading to increased flower drop and poor fruit setting. Several physiological attributes, such as membrane integrity, relative water content and ion/osmotic homeostasis, are disrupted under HT stress. In 21$^{st}$-century agriculture, HT stress has become a significant concern for plant breeders and horticulturists for sustaining the growth and productivity of crop plants under HT stress conditions. Therefore, this chapter reviews HT stress's global and current impact on agriculture. In addition, what preventive measures can be adopted to overcome the adverse effect of HT stress on plants has been critically discussed. Further, how recent advancements in OMICS and biotechnological tools can be exploited for breeding HT tolerance plants.



**Keywords**: HT stress, Growth, Productivity, Phytohormone, Technological interventions


## Introduction

High temperature (HT) stress has become one of the most threatening climate adversities, severely affecting crop phenology, productivity and livelihoods of people around the globe (Pospíšil 2016). Correspondingly, researchers have contemplated that a progressive increase in the earth's air temperature via anthropogenic activities could trigger global warming, thus substantially affecting agricultural produce (Lavania et al. 2015). Frequent climate change has led to a drastic increase in heat waves that have profoundly affected the earth's climatic patterns and global warming (Pospíšil 2016). Furthermore, climate change and increasing heat waves have also altered the earth's geography by increasing deserts and dry and arid land due to frequent drought episodes (Lamers et al. 2020). Recently, Intergovernmental Panel on Climate Change (IPCC) have warned that an increase in air temperature could lead to irregular rainfall, soil salinity/ drought, and $CO_2$ emission (Jehn et al., 2022). Consequently, it has become imperative to delve deeper into the plant-temperature interactive process that may help find novel mechanisms to prevent crop death and agricultural loss (Lavania et al. 2015).

Being sessile, plants are constantly exposed to various biotic and abiotic stresses that severely affect their growth and metabolism (Niu and Xiang, 2018). HT stress is the most devastating abiotic stress that causes huge crop loss in various parts of the world, thus hampering agricultural productivity and global food security (Szymańska et al. 2017). One of the early reverberations of HT stress on plants is the disruption of homeostasis, i.e., ion and osmotic homeostasis. This disruption of homeostasis occurs at both cellular and whole plant levels leading to cellular damage and cell death (Lamers et al., 2020). Therefore, to survive under abiotic stress plant must exhibit three related activities (i) damage must be prevented or alleviated (ii) homeostasis must be re-established in a new and stressful environment (iii) growth must resume albeit at a slow rate (Niu and Xiang 2018).

HT stress instigates detrimental effects on plants due to increased oxidative damage of cells and tissues as provoked by the excess generation of reactive oxygen species (ROS). ROS such as hydrogen peroxide, singlet oxygen and superoxide anion is most commonly known to induce oxidative damage in plants when their levels exceed the threshold limit (Awasthi et al. 2015). These ROS act as signaling molecules at a low level, thus modulating signal transduction pathways under stress conditions. When their level reaches beyond the cell's antioxidative capacity, they can affect many cellular processes, affecting membrane fluidity and leading to cell death and apoptosis (Pospíšil 2016). At a molecular level, HT stress alters expression of stress-responsive genes, regulatory proteins, and osmoprotectants such as proline and glycine betaine (Awasthi et al. 2015).

Therefore, developing HT stress-tolerant crop plants is the need of the hour. However, several factors, limits and perse., are the significant challenges to the plant science community that prevent the development of new HT stress-tolerant cultivars such as HT stress duration, plant type and several other environmental factors such as humidity, heat waves which in conjugation with HT stress represent a significant bottleneck for plant scientists (Rykaczewska 2015). In addition, researchers are yet to identify genes/proteins responsible for imparting HT stress tolerance in plants. Since HT stress tolerance is a quantitative trait governed by multiple genes, linking genes to HT tolerant trait is a rigorous process involving a series of hybridisation experiments using HT tolerant and HT susceptible plants. Plant scientists are exploiting state-of-the-art techniques such as system and synthetic biology tools and trying to unravel how plants can withstand climate adversities. This chapter will discuss HT stress's global and current impact on essential crop plants. Further, an up-to-date perspective of how HT stress is becoming an imminent threat to agriculture, livestock and humankind is comprehensively described. Lastly, strategies for enhancing HT stress tolerance in plants using phytohormones as an external application are also discussed.

## Global scenario of HT stress

Due to the global oceans' vast size and heat capacity, a tremendous amount of heat energy is required to raise the earth's surface temperature even by a smaller fraction (Li et al. 2021). According to the reports from NASA, the earth's surface temperature has increased to roughly 2°C since the pre-industrial era (1880-1900), which may not seem much. Still, it represents a compelling increase in the accumulated heat (Zandalinas et al. 2021). These gradually gathered heat provokes regional and seasonal temperature fluctuations causing melting of sea ice and snow cover, and low/excess rainfall, thus significantly altering the habitat pattern of plants and animals (Chaudhry and Sidhu 2022). The drastic change in habitat ranges could benefit some lifeforms. They are expanding their population, while for most of them, it functions by shrinking their population. For example, the temperature rises exponentially in land areas compared to the oceans, and the Arctic is warming faster than other regions (Chaudhry and Sidhu, 2022).

Recently, the global climate report of 2022 from the National Centres for Environment Information (Figure 1), despite of La Nina climate pattern in the tropical Pacific, each month of 2022 was $10^{th}$ warmer than average (Parmesan et al. 2022). Furthermore, their report concluded that 2022 was the sixth warmest year since the global temperature record was incepted in 1880 (Parmesan et al. 2022). Researchers have shown that the earth's surface temperature increase has not been uniform and inferred that most land areas are warming rather than cooling (Masson-Delmotte et al., 2022). NOAA's 2021 annual climate report, the average temperature of land and ocean showed a combined increase of 0.08°C per decade since 1880, whereas since 1981 average rate of growth in the combined average temperature of land and sea is more than twice, i.e., 0.18°C per decade (Setzer and Higham 2022). According to a study, an increase in the earth's surface temperature and the warming of oceans will depend on the rate of carbon dioxide emission and other greenhouse gases (Masson-Delmotte et al. 2022). Today, deforestation, burning fossil fuels and other anthropogenic activities are accumulating $CO_2$ at an exaggerated rate, i.e., about 11 billion metric tons to over 40 billion metric tons to the atmosphere yearly (Setzer and Higham 2022).

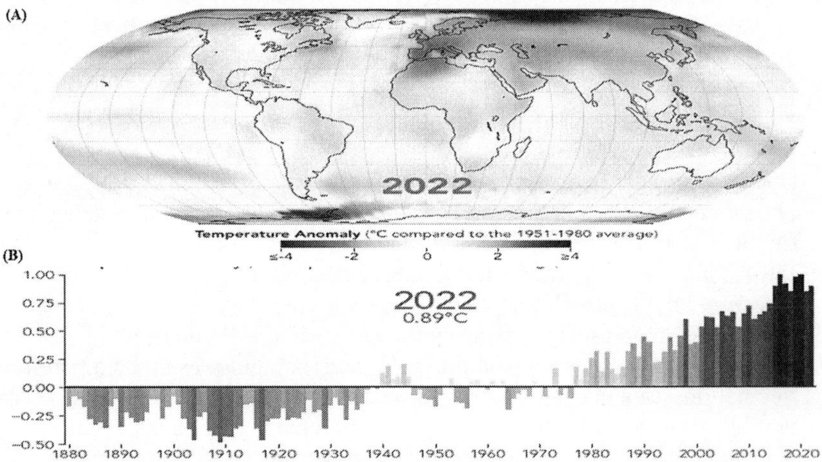


**Figure 1** An overview of global average surface temperature between 1993 and 2022 in degrees Fahrenheit per decade (Adapted from http://www.climate.gov/media/15022)

In the 20th century, we witnessed a rapid increase in the global mean temperature of the earth and oceans. This expansion could be attributed to natural variability and rapid developments of factories, power plants and the use of motor vehicles after World War II (Armstrong McKay et al. 2022). Further, research findings have corroborated that greenhouse accumulates at a slower rate. Still, they remain in the atmosphere for a long time which could also be one of the reasons for a gradual rise in the earth's surface temperature (Armstrong McKay et al. 2022). The temperature fluctuations locally, i.e., within a short period, such as day and night temperatures in summer and winter, don't significantly contribute to global warming.

In contrast, the increase in global warming is mainly dependent on the rise in the earth's surface temperature, which is reflected in terms of sunlight received by the planet and the amount of it radiates back to space (Moran et al. 2008). The amount of solar radiation that emits back to space is proportional to the composition of greenhouse gases in the atmosphere (Zandalinas et al. 2021). Even a single degree increase in the global mean temperature is significant because it takes a massive amount of heat energy warm the ocean and surface of the earth (Zandalinas et al. 2021). The image below depicts the global temperature anomalies in 2022.

## Current scenario of HT stress

The earth's surface temperature is progressively increasing, and heat waves are becoming frequent and intense around the globe. For example., in India, intense HT has been recorded in April and May every year, of which 2022 recorded an unusual increase in the maximum and minimum temperature (He et al. 2022). According to the current Indian metrological report, 2022 witnessed extreme temperatures, i.e., up to 8°C–10.8°C and much lower (60-80%) rainfall (He et al. 2022). Agricultural scientists remember 2022 as one of the most catastrophic years, severely affecting agricultural productivity due to extreme temperatures and low to moderate rainfall (Chand et al., 2022). In recent years, Indian agriculture is becoming increasingly vulnerable to climate extremes mainly due to temperature fluctuations and low rainfall that have significantly impacted the growth and productivity of crop plants (Chand et al. 2022). The researchers delve deeper into the action of HT stress on decreasing the productivity of crop plants. They identified that HT stress resulted in sunburn, lipid peroxidation and reduced moisture content resulting in flower drop and low fruit setting in economically important crops (Chaudhry and Sidhu 2022). In addition, the review report of IPCC convoluted that climate change is happening faster than expected, and its effect can be seen globally as the production of major staple crops is steadily declining (Shekhawat et al. 2022).

In India, a temperature rise significantly alters rainfall frequency, timing and magnitude, affecting relative humidity, heat waves and water availability to agricultural lands (Singh et al., 2022). Indian scientists have elucidated heat waves' magnitude by combining heatwaves and drought dossiers and employing a standardised precipitation index (SPI) (Shekhawat et al. 2022). They concluded that the area affected by high temperature had increased to nearly 4 per cent since 2010, a significant concern for agriculture and allied sectors (Rivero et al. 2022). The increasing temperature or heat waves in the inevitable part of India are primarily due to the persistent warm anticyclone-inducing synoptic situation favouring heat waves (Singh et al. 2022). It could also be due to the absence of adequate moisture in the upper atmosphere, or there is practically no cloud in the sky (Ding and Yang, 2022). Low rainfall can also cause the drying of soil, due to which intense solar radiations can sporadically warm the surface of the ground, thus exaggerating the soil temperature (Rivero et al., 2022).

Still, the World Metrological Organization (WMO) has yet to adopt a standard definition for heat waves. Still, according to IMD, heat waves become severe if the temperature of a particular region becomes 5-6°C higher than usual. Understanding the gravity of the situation, IMD has portrayed those extreme climatic conditions like the present one could become more frequent in the coming years due to climate change. Therefore, concerted efforts are required to develop new technologies that can map variations in climatic conditions in real-time, thereby allowing the implementation of protective measures for improving crop growth and productivity under harsh environments. Furthermore, researchers are also rewiring plants' innate immune systems by growing them in an artificial environment that mimics the natural environment, which may help identify new genes/proteins that help them withstand climate extremes.

## Impact of HT Stress on 21st-century Agriculture

High-temperature stress, particularly in March and April, profoundly affects crop plants' growth and reproduction, thus limiting agricultural productivity and global food security (Sun et al. 2019). The first physical impact of HT is specifically attributed to the loss of water from the soil or within the plants (Table 1) (Herold et al. 2018). Water loss induces significant changes in plants' ion/osmotic homeostasis, thus affecting their oxidative phosphorylation and photophosphorylation (Mall et al. 2017). In extreme cases, the low relative water content may also induce chlorophyll degradation due to the peroxidation of membranes and electrolytic leakage (Lobell et al. 2015). The generation of oxidative stress is also an essential physiological impact of HT, affecting many cellular processes in plants, such as cell divisions and elongation (Adhikari et al. 2015). The generated

**Table 1**: High-temperature stress and its impact on agriculture in the current scenario

| Topic | Information | References |
|---|---|---|
| Definition of High-Temperature Stress | High-temperature stress in plants refers to the exposure of crops to elevated temperatures beyond their optimal range, leading to physiological and metabolic disruptions. | IPCC. (2014) |
| Impact on Crop Growth and Yield | High temperatures can negatively affect various aspects of plant growth and development, leading to reduced crop yield and quality. The impacts include: | Challinor et al. (2014) |
| | Decreased photosynthesis and carbon assimilation | Lobell and Gourdji (2012) |
| | Impaired reproductive development and reduced pollination | Schlenker and Roberts (2009) |
| | Increased water stress and reduced water-use efficiency | Hatfield et al. (2011) |
| | Altered nutrient uptake and nutrient imbalances | Raja et al. (2018) |
| | Enhanced susceptibility to pests, diseases, and weeds | Lemoine (2015) |
| Strategies to Mitigate High-Temperature Stress | To mitigate the impact of high-temperature stress on agriculture, several strategies can be employed, such as: | Bita and Gerats (2013) |
| | Developing heat-tolerant crop varieties through breeding and genetic engineering | Prasad and Djanaguiraman (2014) |
| | Implementing agronomic practices, such as adjusting planting dates and irrigation scheduling | Mohanty et al. (2013) |
| | Adoption of conservation agriculture practices to improve soil moisture retention | Lal (2015) |
| | Enhancing crop resilience through foliar applications of stress-mitigating compounds | Farooq et al. (2011) |

oxidative stress provokes stomatal closure, thus affecting the $CO_2$ diffusion efficiency and stimulating photorespiration leading to cell death and apoptosis (Jin et al. 2016). In addition, HT stress also negatively modulates the transpiration rate, thus affecting the internal continuum of plants due to reduced cooling and increased temperature in cellular compartments (Sun et al. 2019).

Several studies have proclaimed that HT-induced oxidative stress is often associated with intense heat waves and high water use efficiency. C3 plants generally can adapt themselves to temperature variations. However, temperature above a certain threshold negatively affects photosynthesis and respiration. CAM plants can also acclimate to extreme environments as they can hibernate in broad daylight and activate their photosynthesis at night. C4 plants can tolerate HT stress; therefore, their growth and productivity are considerably higher than C3 and CAM plants (Lobell et al. 2015). In addition, livestock productivity is also greatly influenced by the HT environment (Malhi et al. 2021). According to the reports, the ambient environment for livestock is when the thermal heat index (THI) is below or equal to 72 and THI above 85, drastically affecting their feed intake and productivity (Adhikari et al. 2015). Due to lower feed intake, the livestock develops severe implications such as reduced libido, fertility and embryonic survival (Belhadj Slimen et al. 2016). HT stress also severely affects their metabolism by disrupting ion homeostasis due to a rapid decrease in $NA^+$ and $K^+$ (Belhadj Slimen et al. 2016).

HT stress negatively impacts poultry farming by causing broilers discomfort so they can no longer reproduce (Polsky et al. 2017). HT greatly influences their growth, behaviour and immune response by hampering their feed conversion and relative weight gain (Polsky et al. 2017). A recently published report signifies that fluctuation in ambient temperature provokes birds to reduce their feed consumption by 5-7% for every 1°C rise in ambient temperature (Das et al. 2016). The temperature fluctuations make it difficult for birds to balance heat production and heat loss, reducing their dietary digestibility and leading to protein loss and decreased calcium levels (Das et al. 2016). HT stress also affects the hen's reproducibility, thus affecting the production of quality eggs (Yavuz et al. 2023). Likewise, an increase in ocean temperature harms fisheries and aquaculture due to variations in sea surface temperature, circulation patterns and increased acidification via anthropogenic activities (Polsky et al. 2017). Recently, findings have confirmed that a global rise in temperature will harm tropical and subtropical ecosystems (Polsky et al. 2017). Drastic temperature shifts will increase the sea level due to the melting the polar ice cap and glaciers, thus impacting mangrove forests and marine fish nursery grounds (Belhadj Slimen et al. 2016). High sea surface temperature will also directly affect swimming and the physiological properties of fish by lowering the level of dissolved oxygen (Yavuz et al. 2023). The low level of dissolved oxygen could be attributed to the effect of HT on stratification leading to rapid expansion in an algal bloom.

## Traditional strategies for the management of HT stress

HT in different parts of the world has triggered massive heat waves severely affecting the growth and productivity of crop plants (Wen et al. 2019). According to a report, major wheat-growing regions in India experienced a 3-4°C increase in average temperature as compared to the previous year, thus hampering both the quality and quantity of grains (Tiwari et al. 2017). In India, wheat is usually grown in the month of February-March in an area of about 31 million hectares, and the crops usually experience HT stress during the $4^{th}$ weeks of March and $1^{st}$ week of April (Yadav et al. 2022). The farmers use various protective measures to protect their crops from HT stress-induced oxidative damage. First, exploitation of heat-tolerant varieties to develop new hybrids capable of withstanding HT. Technologies such as residue management, direct seeding, and a spray of $KNO_3$ (0.5%) at anthesis stages can minimise yield losses during HT stress conditions (Singh et al. 2020). Repeated irrigation from time-time can alleviate HT stress-induced oxidative damages by maintaining ion/osmotic homeostasis in plants (Janni et al. 2020). HT stress significantly decreases the fruit setting in plants due to poor pollination leading to flower drop (Godoy et al. 2021).

Researchers have used drip irrigation along with organic mulch to prevent HT stress-induced oxidative damage in plants (Mendonca et al. 2021). Mulching helps retain prolonged soil moisture under HT stress, and foliar spray with 2-4% kaolin will prevent water loss in the form of transpiration, thus improving the physiological process under temperature extremes (Mendonca et al. 2021). In tomato plants, researchers have used drip irrigation twice daily, followed by an organic mulch of 5-8 cm thickness to maintain a friendly environment in the soil. In addition, some researchers also recommend planting maise plants to protect tomato plants from heat waves (Mendonca et al. 2021). Further, external applications of salicylic acid and nitric oxide have also been proven to alleviate HT stress-induced oxidative damage in tomato plants (Zhang et al. 2020). In radish, using a shading net with frequent drip irrigation maintain soil moisture and significantly boosts growth and productivity under HT stress conditions (Matamoros et al. 2022).

The techniques mentioned above are traditional ways to minimise HT stress's negative effect on crop plants. These techniques stimulate short-term adaptive mechanisms in plants by modulating their innate immune response, thus improving their resilience under climate extremes (Matamoros et al. 2022). However, to develop crop plants resilient to HT stress, researchers are exploiting molecular biology tools in conjunction with OMICS techniques to identify genes/proteins conferring HT stress tolerance (Table 2) (Bhardwaj et al. 2021). The identified genes are then used to develop transgenic plants that can withstand harsh environmental cues without affecting their growth and productivity (Janni et al. 2020). Since the cultivation of transgenic plants is limited and, in many countries, it is not approved by the government, enhancing HT stress tolerance via alternative means is the only viable option (Janni et al. 2020). Recently, exogenous applications of various phytohormones and osmoprotectants have been shown to modulate physiological and biochemical processes in plants exposed to abiotic and biotic stresses. A brief about the role of these phytohormones and osmoprotectants is described in the following sections.

**Table 2** Traditional and modern strategies for the management of high-temperature stress in plants

| Strategy | Description | References |
|---|---|---|
| **Traditional Strategies** | | |
| Shade Provision | Providing temporary or permanent shade structures to reduce direct exposure of plants to high temperatures. | Nadeem et al. (2018) |
| Mulching | Applying organic or inorganic mulch materials around plants to reduce soil temperature, conserve moisture, and suppress weed growth. | Farooq et al. (2011) |
| Traditional Crop Selection and Management | Selecting and cultivating crop varieties that are known to tolerate or adapt better to high-temperature conditions. | Mittler (2006) |

| | | |
|---|---|---|
| **Modern Strategies** | | |
| Genetic Engineering | Developing transgenic crops with improved heat tolerance by introducing genes that regulate heat shock proteins or antioxidants. | Wahid et al. (2007) |
| Marker-Assisted Breeding | Using molecular markers to identify and select plants with desired heat tolerance traits during the breeding process. | Najeeb et al. (2020) |
| Exogenous Application of Plant Growth Regulators | Applying plant growth regulators, such as brassinosteroids or salicylic acid, to enhance heat stress tolerance in plants. | Hasanuzzaman et al. (2012, 2013) |
| Precision Irrigation | Implementing irrigation systems that provide water directly to the root zone, optimizing water use efficiency and reducing heat stress. | Singh et al. (2014) |
| Protective Foliar Sprays | Applying foliar sprays containing natural or synthetic compounds to enhance the plant's tolerance to high-temperature stress. | Joshi et al. (2007) |
| Soil Amendments | Incorporating soil amendments, such as organic matter or biochar, to improve soil structure, water holding capacity, and nutrient availability, which can mitigate high-temperature stress effects. | Asli and Neumann (2010) |

## Advanced strategies for management of HT stress

### Phytohormones and osmoprotectants conferring HT stress tolerance

Certain phytohormones, signaling molecules and osmoprotectants have the potential to counteract the adverse effect of abiotic stress-induced oxidative damage in plants (Wu et al. 2019). Recent decades have witnessed an exponential increase in the application of these phytohormones, signaling molecules and osmoprotectants due to their stress ameliorative and growth-promoting properties (Rai et al. 2018, 2020). Phytohormones such as salicylic acid (SA), jasmonic acid (JA), Abscisic acid (ABA) and nitric oxide (NO) as signaling molecules and osmoprotectants such as proline (Pro) or glycine (GB) betaine have been shown to stimulate a plethora of physiological and biochemical process in plants exposed to HT stress (Rai et al. 2020a). In a study, exogenous application of Pro (20 mM) and GB (20 mM) decreases the impact of HT stress-induced oxidative damage in sugarcane plants (Tiwari et al. 2017). Both Pro and GB treatment could have modulated the activity of antioxidative enzymes, thereby maintaining the internal homeostasis of plants under HT stress conditions (Rai et al. 2018). Similarly, another study also confirmed the above notion where the external application of Pro and GB modulate the level of soluble sugars, $Na^+$, and $K^{+,}$ thus effectively maintaining membrane integrity in HT-stressed plants (Tiwari et al. 2017). Exogenous application of Pro in HT-stressed chickpea plants controlled the generation of MDA and $H_2O_2$ levels due to enhanced activity of enzymatic and non-enzymatic antioxidants (Parankusam et al. 2017). Further, in their study, they identified a 20-30% increase in shoot length in Pro-treated chickpea plants as compared to non-treated ones, which is indicative of growth promotive and stress ameliorative properties of Pro (El-Beltagi et al. 2020).

SA is a well-acclaimed phytohormone whose stress-ameliorative nature has long been diagnosed (Kaur et al. 2022). It is a phenolic compound, and its controlled application has been reported to have a favourable implication on the growth and immune system of plants exposed to harsh environmental conditions (Rai et al. 2018). External application of SA (0.1 mM) compellingly reduced the generation of thiobarbituric acid reactive substance (TBARS), thus strengthening membrane integrity in grape plants exposed to HT stress (Rai et al. 2020). Likewise, researchers have noted improvement in photosystem II (PSII) efficiency in grapevines upon treating them with 0.1 mM of SA. This effect could be due to SA-mediated modulation of defence signaling genes which could have regulated the activity of RUBISCO, thus maintaining optimum photosynthesis in grapevines under HT stress (Kaur et al. 2022). Similarly, another study reported that treating rice seedlings with 0.5 mM SA significantly altered the effect of HT-induced oxidative damage by contriving ROS generation and its immediate impacts, such as membrane damage and electrolytic leakage (Rai et al. 2020a). Further, researchers identified the molecular notion behind the SA effect. They concluded that it was due to the SA-mediated regulation of stress-responsive genes, particularly heat shock proteins (HSPs), that rendered HT tolerance in rice seedlings (Rai et al. 2020).

ABA, which is widely known as the stress hormone, is also widely exploited externally for improving the growth and productivity of crop plants exposed to environmental cues (Chen et al. 2020). Correspondingly, researchers have demarcated that ABA has been extensively used for imparting HT stress tolerance in plants (Kumar et al. 2019). In chickpea plants, foliar application of ABA (2.5 μM) progressively improved the growth of the seedlings at 40-45°C (Pang et al. 2017). This improvement could be due to ABA-mediated enhanced production of endogenous proline that protected the membrane integrity of seedlings growing under HT stress. In addition, ABA application could also have modulated ROS generation and enzymatic antioxidants, which ultimately have improved their innate immune response (Chu et al. 2018). Conversely, researchers applied fluridone (ABA inhibitor) to counter the stress ameliorative effect of ABA on chickpea plants (Kaur et al. 2019). They identified that exogenous application of fluridone enhances HT stress-induced oxidative damage by stimulating ROS generation, lowering the activities of enzymatic antioxidants and decreasing the level of proline (Kaur et al. 2019). Their study confirmed that the ABA was responsible for improving chickpea plants' growth and productivity under HT stress.

JA is another phytohormone extensively employed to induce plants' abiotic and biotic stress tolerance. Exogenous application of JA (50 μM) to the HT-stressed chickpea enhanced their growth and productivity (Deepika et al. 2022). This protection could be due to the JA-mediated activation of defence signaling pathways that modulated the antioxidant defence system in grapevines exposed to HT (Wang et al. 2020). Similarly, the external application of JA improved fatty acid composition in the membrane of *B. juncea* plants, thus improving their membrane integrity and ion homeostasis under HT stress (Wang et al. 2020). This effect of JA could be due to improved activities of CAT, SOD and APX that may have contrived the level of ROS below the threshold level where they have been effectively used as singling molecules leading to improve growth (Deepika et al. 2022).

NO is a gaseous signaling molecule implicated in the regulation of various physiological and biochemical processes in plants (Fancy et al. 2017). In plants most commonly used NO donor is sodium nitroprusside (SNP) which, when applied at lower concentrations, promotes growth and improved tolerance in plants under abiotic and biotic stress conditions (Rai et al. 2018, 2020). Several researchers have exploited NO donor SNP at low concentrations (2.5 μM), for example, in hyacinth bean, tomato, soybean, chickpea etc. and have well corroborated the growth-promoting effect of SNP (Rai et al. 2018, 2020). In a study, foliar application of SNP (1 and 2.5 μM) significantly improved HT stress tolerance in the callus of *P. communis*. On further investigation, the researchers observed that the application of SNP has tremendously reduced the ROS generation and electrolytic leakage, which have significantly improved the callus growth under HT stress (Corpas et al. 2011).

Similarly, exogenous application SNP have extensively improved the flowering and pod setting in hyacinth bean plants exposed to HT stress. The improved pod setting in the hyacinth bean plants was due to NO-mediated activation of DNA demethylation, which has activated the expression of various downstream stress-responsive genes (Oz et al. 2015). In addition, the SNP-treated plants also showed improved physiological attributes such as lipid peroxidation, relative water content and electrolytic leakage compared to non-treated plants (Oz et al. 2015). Likewise, foliar application of SNP (0.5 mM) improved the level of non-enzymatic antioxidants such as ascorbate and glutathione, thereby improving the growth of hydroponically grown wheat seedlings under HT stress (Parankusam et al. 2017). Various researchers have used the combined application of SA, NO, JA, or ABA to stimulate HT stress tolerance in plants. Their observations mainly postulated that exogenous application of the phytohormones mentioned above could improve stress tolerance and modulate their adaptation under stress conditions if applied at low concentrations. They could have detrimental effects at high concentrations and negatively impact their growth and productivity. So, obtaining a lethal dose ($LD_{50}$) for respective phytohormones and test plants is better than performing external applications to get desired results.

## Biotechnological interventions for developing strategies for enhancing HT stress tolerance in plants

Advancements in OMICS approaches have delineated novel mechanistic ways to decipher new signaling pathways that modulate stress tolerance in crop plants (Figure 2). A comprehensive understanding of the transcriptional, translational and post-translational process has provided an in-depth knowledge of genomics, proteomics and metabolomics of crop plants that have resulted in the identification of new genes/proteins responsible for stress tolerance (Hasanuzzaman et al. 2013). The discovery

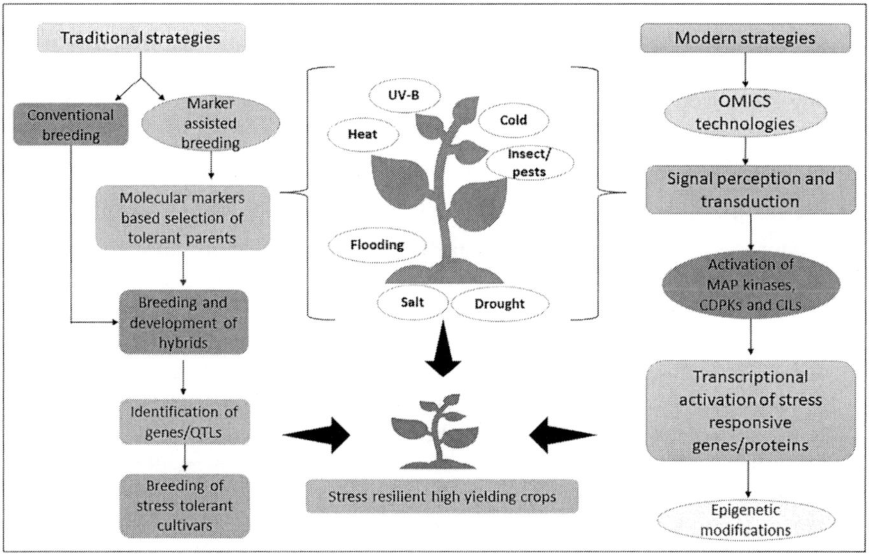


**Figure 2** Different conventional and molecular biology (OMICS) tools that can be used to improve heat stress tolerance in plants.

of several small RNAs, such as small interfering RNA (siRNA) and microRNA (miRNA), has also paved the way to confer abiotic stress tolerance in plants. These small RNAs are specifically involved in the regulation of specific genes either by creating gene knockout or mutant (Bagati et al. 2021). Furthermore, researchers have also identified some miRNA-resistant target genes which can be exploited for the generation of transgenic plants that will have better adaptability under stress conditions (Marco et al. 2015). The engineered plants will be capable of overcoming various transcriptional and post-translational hurdles thus ensuring improved physiological and biochemical performance of crop plants (Zhang et al. 2000). Microarray is also one of the most powerful tools to decipher the expression pattern of several genes simultaneously (Hasanuzzaman et al. 2013). This technique has been extensively employed for identifying genes responsible for conferring biotic/abiotic stress tolerance in plants (Shinozaki and Yamaguchi-Shinozaki 2007). Microarray facilitated the identification of 150 cDNA transcripts in *N. tabacum* exposed to drought as well as heat stress (Seki et al. 2009). Likewise, researchers have also successfully used the microarray technique and identified several transcripts in *Arabidopsis thaliana* that were being upregulated under heat and drought stress (Li et al. 2016).

Recently, researchers have exploited the microarray technique to expand current knowledge of HT stress tolerance by identifying the role of HSPs (Hasanuzzaman et al. 2013). While studying the microarray pattern in HT-stressed Arabidopsis plants which were deficient in the ascorbate peroxidase (APX) gene, they observed that certain HSPs were being upregulated (Asthir 2015).

HSPs are those proteins which are encoded by heat-inducible genes commonly known as heat shock factors (HSFs) which are only expressed under stress conditions predominantly under HT stress (Yadav et al. 2020). HSPs are highly tolerant to high temperatures and upregulated upon binding to conserve cis-regulatory elements known as heat shock (HSEs). In various crop plants, HSFs are constitutively expressed and, upon binding to HSEs, regulate the activity of transcriptional components, thus modulating gene expression under stress conditions (Yadav et al. 2020). In addition, studies have indicated several HSPs are heat-inducible, and their effect on activating the transcription of nearby genes depends mainly on the duration and intensity of HT stress (Yang et al. 2021). Researchers have also developed several transgenic plants where HSP expression has been modulated either by creating gene knockout or mutant (Hemantaranjan et al. 2018). For example, in Arabidopsis, researchers successfully altered the expression of heat shock protein *AtHSF1* and the resulting plants suffered extensive damage under HT stress as transcription of other stress-responsive genes were repressed due to downregulation of *AtHSF1* (Hasanuzzaman et al. 2013). HSPs also function as molecular chaperones which help maintain the structure of proteins under stress conditions by preventing them from being denatured (Hemantaranjan et al. 2018). This notion of HSPs has been validated in various crop plants such as tobacco, rice and legumes where constitutive expression of HSPs was made using CaMV35 S promoter, which enhanced the thermotolerance of these plants by activating stress-responsive proteins and other metabolites (Yang et al. 2021). The biotechnological interventions can be used for the genetic improvement of crop plants by exploiting them either in a breeding program or by developing transgenic plants to improve growth and agricultural productivity.

## Conclusion

HT stress has become a daunting challenge worldwide due to its severe repercussion on crop growth and productivity. The severity of HT stress depends upon the intensity and duration of the HT stress condition and the plant's age and morphology. The maximum and minimum temperatures are expected to increase gradually owing to the increase in greenhouse gases, thus initiating the chain of global warming. Several studies/research have been conducted, and initiatives have been taken to understand the response of plants exposed to HT stress. Nonetheless, no effective measures or mechanisms have been identified to enhance plant thermotolerance. Researchers can improve short-term HT stress tolerance in plants by applying certain phytohormones, signaling molecules and osmoprotectants. Alternatively, researchers could also improve HT stress tolerance in plants by altering the time of sowing, irrigation and use of wild cultivars. However, their productivity could not be achieved at to desired level. The understanding of molecular mechanisms governing HT stress tolerance seems far-fetched. Therefore, extensive research is required to develop stress-tolerant cultivars either by conventional/molecular breeding approaches or by manipulating the genomic region of the desired crops to create stress memory. The development of stress memory will enable the plant to prepare for the upcoming situation by synthesising effective metabolites or enhancing stress-responsive genes' expression. Stress memory in plants could be achieved by rewiring crop plants at the molecular level using OMICS approaches and exploitation of which could engineer super plants which can withstand the harsh environment without affecting their productivity, thus ensuring global food security in the era of climate change.

## References

Adhikari, U., Nejadhashemi, A.P. and Woznicki, S.A., 2015. Climate change and eastern Africa: a review of impact on major crops. Food and Energy Security, 4(2), pp.110-132.

Armstrong McKay, D.I., Staal, A., Abrams, J.F., Winkelmann, R., Sakschewski, B., Loriani, S., Fetzer, I., Cornell, S.E., Rockström, J. and Lenton, T.M., 2022. Exceeding 1.5 C global warming could trigger multiple climate tipping points. Science, 377(6611), p.eabn7950.

Asli, S., & Neumann, P. M. (2010). Rhizosphere humic acid interacts with root cell walls to reduce hydraulic conductivity and plant development. Plant and Soil, 327(1-2), 133-144.

Asthir, B., 2015. Protective mechanisms of heat tolerance in crop plants. Journal of Plant Interactions, 10(1), pp.202-210.

Awasthi, R., Bhandari, K. and Nayyar, H., 2015. Temperature stress and redox homeostasis in agricultural crops. Frontiers in Environmental Science, 3, p.11.

Bagati, S., Rai, G.K., Bhadwal, D. and Berwal, M.K., 2021. Biotechnological approaches for the development of heat stress tolerance in crop plants. In Abiotic Stress Tolerance Mechanisms in Plants (pp. 85-126). CRC Press.

Belhadj Slimen, I., Najar, T., Ghram, A. and Abdrrabba, M.J.O.A.P., 2016. Heat stress effects on livestock: molecular, cellular and metabolic aspects, a review. Journal of animal physiology and animal nutrition, 100(3), pp.401-412.

Bhardwaj, A., Devi, P., Chaudhary, S., Rani, A., Jha, U.C., Kumar, S., Bindumadhava, H., Prasad, P.V., Sharma, K.D., Siddique, K.H. and Nayyar, H., 2021. 'Omics' approaches in developing combined drought and heat tolerance in food crops. Plant Cell Reports, pp.1-41.

Bita, C. E., & Gerats, T. (2013). Plant tolerance to high temperature in a changing environment: Scientific fundamentals and production of heat stress-tolerant crops. Frontiers in Plant Science, 4, 273.

Challinor, A. J., et al. (2014). A meta-analysis of crop yield under climate change and adaptation. Nature Climate Change, 4(4), 287-291.

Chand, S.S., Walsh, K.J., Camargo, S.J., Kossin, J.P., Tory, K.J., Wehner, M.F., Chan, J.C., Klotzbach, P.J., Dowdy, A.J., Bell, S.S. and Ramsay, H.A., 2022. Declining tropical cyclone frequency under global warming. Nature Climate Change, 12(7), pp.655-661.

Chaudhry, S. and Sidhu, G.P.S., 2022. Climate change regulated abiotic stress mechanisms in plants: A comprehensive review. Plant Cell Reports, 41(1), pp.1-31.

Chen, K., Li, G.J., Bressan, R.A., Song, C.P., Zhu, J.K. and Zhao, Y., 2020. Abscisic acid dynamics, signaling, and functions in plants. Journal of integrative plant biology, 62(1), pp.25-54.

Chu, H.D., Nguyen, K.H., Watanabe, Y., Le, D.T., Pham, T.L.T., Mochida, K. and Tran, L.S.P., 2018. Identification, structural characterization and gene expression analysis of members of the Nuclear Factor-Y family in chickpea (*Cicer arietinum* L.) under dehydration and abscisic acid treatments. International journal of molecular sciences, 19(11), p.3290.

Corpas, F.J., Leterrier, M., Valderrama, R., Airaki, M., Chaki, M., Palma, J.M. and Barroso, J.B., 2011. Nitric oxide imbalance provokes a nitrosative response in plants under abiotic stress. Plant Science, 181(5), pp.604-611.

Das, R., Sailo, L., Verma, N., Bharti, P., Saikia, J., Imtiwati, K.R. and Kumar, R., 2016. Impact of heat stress on health and performance of dairy animals: A review. Veterinary world, 9(3), pp.260-268.

Deepika, D., Jonwal, S., Mali, K.V., Sinha, A.K. and Singh, A., 2022. Molecular analysis indicates the involvement of Jasmonic acid biosynthesis pathway in low-potassium ($K^+$) stress response and development in chickpea (*Cicer arietinum*). Environmental and Experimental Botany, 194, p.104753.
Ding, Y. and Yang, S., 2022. Surviving and thriving: How plants perceive and respond to temperature stress. Developmental Cell.
El-Beltagi, H.S., Mohamed, H.I. and Sofy, M.R., 2020. Role of ascorbic acid, glutathione and proline applied as singly or in sequence combination in improving chickpea plant through physiological change and antioxidant defense under different levels of irrigation intervals. Molecules, 25(7), p.1702.
Fancy, N.N., Bahlmann, A.K. and Loake, G.J., 2017. Nitric oxide function in plant abiotic stress. Plant, Cell & Environment, 40(4), pp.462-472.
Farooq, M., Bramley, H., Palta, J.A. and Siddique, K.H., 2011. Heat stress in wheat during reproductive and grain-filling phases. Critical Reviews in Plant Sciences, 30(6), pp.491-507.
Hasanuzzaman, M., Hossain, M.A., da Silva, J.A.T. and Fujita, M., 2012. Plant response and tolerance to abiotic oxidative stress: antioxidant defense is a key factor. Crop stress and its management: perspectives and strategies, pp.261-315.
Hasanuzzaman, M., Nahar, K., Alam, M.M., Roychowdhury, R. and Fujita, M., 2013. Physiological, biochemical, and molecular mechanisms of heat stress tolerance in plants. International journal of molecular sciences, 14(5), pp.9643-9684.
Hatfield, J. L., et al. (2011). Climate impacts on agriculture: Implications for crop production. Agronomy Journal, 103(2), 351-370.
He, W., Zhang, L. and Yuan, C., 2022. Future air temperature projection in high-density tropical cities based on global climate change and urbanisation–a study in Singapore. Urban Climate, 42, p.101115.
Hemantaranjan, A., Malik, C.P. and Bhanu, A.N., 2018. Physiology of heat stress and tolerance mechanisms—an overview. J Plant Sci Res, 33(1), pp.55-68.
Herold, N., Ekström, M., Kala, J., Goldie, J. and Evans, J.P., 2018. Australian climate extremes in the 21st century according to a regional climate model ensemble: Implications for health and agriculture. Weather and climate extremes, 20, pp.54-68.
IPCC. (2014). Climate Change 2014: Impacts, Adaptation, and Vulnerability. Part A: Global and Sectoral Aspects. Contribution of Working Group II to the Fifth Assessment Report of the Intergovernmental Panel on Climate Change. Cambridge University Press.

Janni, M., Gullì, M., Maestri, E., Marmiroli, M., Valliyodan, B., Nguyen, H.T. and Marmiroli, N., 2020. Molecular and genetic bases of heat stress responses in crop plants and breeding for increased resilience and productivity. Journal of Experimental Botany, 71(13), pp.3780-3802.

Jehn, F.U., Kemp, L., Ilin, E., Funk, C., Wang, J.R. and Breuer, L., 2022. Focus of the IPCC assessment reports has shifted to lower temperatures. Earth's Future, 10(5), p.e2022EF002876.

Jin, Z., Zhuang, Q., Tan, Z., Dukes, J.S., Zheng, B. and Melillo, J.M., 2016. Do maise models capture the impacts of heat and drought stresses on yield? Using algorithm ensembles to identify successful approaches. Global Change Biology, 22(9), pp.3112-3126.

Joshi, A.K., Mishra, B., Chatrath, R., Ortiz Ferrara, G. and Singh, R.P., 2007. Wheat improvement in India: present status, emerging challenges and future prospects. Euphytica, 157, pp.431-446.

Kaur, D., Grewal, S.K., Kaur, J. and Singh, S., 2019. Differential organ specific protein profiling in chickpea cultivars under water deficit condition. Journal of Food Legumes, 32(4), pp.221-226.

Kaur, H., Hussain, S.J., Kaur, G., Poor, P., Alamri, S., Siddiqui, M.H. and Khan, M.I.R., 2022. Salicylic acid improves nitrogen fixation, growth, yield and antioxidant defence mechanisms in chickpea genotypes under salt stress. Journal of Plant Growth Regulation, 41(5), pp.2034-2047.

Kumar, M., Kesawat, M.S., Ali, A., Lee, S.C., Gill, S.S. and Kim, H.U., 2019. Integration of abscisic acid signaling with other signaling pathways in plant stress responses and development. Plants, 8(12), p.592.

Lal, R. (2015). Restoring soil quality to mitigate soil degradation. Sustainability, 7(5), 5875-5895.

Lamers, J., Van Der Meer, T. and Testerink, C., 2020. How plants sense and respond to stressful environments. Plant Physiology, 182(4), pp.1624-1635.

Lavania, D., Dhingra, A., Siddiqui, M.H., Al-Whaibi, M.H. and Grover, A., 2015. Current status of the production of high temperature tolerant transgenic crops for cultivation in warmer climates. Plant Physiology and Biochemistry, 86, pp.100-108.

Lemoine, N., 2015. The Effects of Climate Warming on Plant-Herbivore Interactions.

Li, H., Wang, Y., Wang, Z., Guo, X., Wang, F., Xia, X.J., Zhou, J., Shi, K., Yu, J.Q. and Zhou, Y.H., 2016. Microarray and genetic analysis reveals that csa-miR159b plays a critical role in abscisic acid-mediated heat tolerance in grafted cucumber plants. Plant, cell & environment, 39(8), pp.1790-1804.

Li, N., Euring, D., Cha, J.Y., Lin, Z., Lu, M., Huang, L.J. and Kim, W.Y., 2021. Plant hormone-mediated regulation of heat tolerance in response to global climate change. Frontiers in Plant Science, 11, p.627969.

Lobell, D. B., & Gourdji, S. M. (2012). The influence of climate change on global crop productivity. Plant Physiology, 160(4), 1686-1697.

Lobell, D.B., Hammer, G.L., Chenu, K., Zheng, B., McLean, G. and Chapman, S.C., 2015. The shifting influence of drought and heat stress for crops in northeast Australia. Global change biology, 21(11), pp.4115-4127.

Malhi, G.S., Kaur, M. and Kaushik, P., 2021. Impact of climate change on agriculture and its mitigation strategies: A review. Sustainability, 13(3), p.1318.

Mall, R.K., Gupta, A. and Sonkar, G., 2017. Effect of climate change on agricultural crops. In Current developments in biotechnology and bioengineering (pp. 23-46). Elsevier.

Marco, F., Bitrián, M., Carrasco, P., Rajam, M.V., Alcázar, R. and Tiburcio, A.F., 2015. Genetic engineering strategies for abiotic stress tolerance in plants. Plant Biology and Biotechnology: Volume II: Plant Genomics and Biotechnology, pp.579-609.

Masson-Delmotte, V., Zhai, P., Pörtner, H.O., Roberts, D., Skea, J. and Shukla, P.R., 2022. Global Warming of 1.5° C: IPCC Special Report on Impacts of Global Warming of 1.5° C above Pre-industrial Levels in Context of Strengthening Response to Climate Change, Sustainable Development, and Efforts to Eradicate Poverty. Cambridge University Press.

Matamoros, V., Casas, M.E., Mansilla, S., Tadić, Đ., Cañameras, N., Carazo, N., Portugal, J., Piña, B., Díez, S. and Bayona, J.M., 2022. Occurrence of antibiotics in Lettuce (*Lactuca sativa* L.) and Radish (*Raphanus sativus* L.) following organic soil fertilisation under plot-scale conditions: Crop and human health implications. Journal of hazardous materials, 436, p.129044.

Mendonca, S.R., Ávila, M.C.R., Vital, R.G., Evangelista, Z.R., de Carvalho Pontes, N. and dos Reis Nascimento, A., 2021. The effect of different mulching on tomato development and yield. Scientia Horticulturae, 275, p.109657.

Mittler, R. (2006). Abiotic stress, the field environment and stress combination. Trends in Plant Science, 11(1), 15-19.

Mohanty, S., Wassmann, R., Nelson, A., Moya, P. and Jagadish, S.V.K., 2013. Rice and climate change: significance for food security and vulnerability. International Rice Research Institute, 14, pp.1-14.

Moran, D., MacLeod, M., Wall, E., Eory, V., Pajot, G., Matthews, R., McVittie, A., Barnes, A.P., Rees, R.M., Moxey, A.P. and Williams, A., 2008. UK marginal abatement cost curves for the agriculture and land use, land-use change and forestry sectors out to 2022, with qualitative analysis of options to 2050: Final Report to the Committee on Climate Change.

Nadeem, M., Li, J., Wang, M., Shah, L., Lu, S., Wang, X. and Ma, C., 2018. Unraveling field crops sensitivity to heat stress: Mechanisms, approaches, and future prospects. Agronomy, 8(7), p.128.

Najeeb, S., Ali, J., Mahender, A., Pang, Y.L., Zilhas, J., Murugaiyan, V., Vemireddy, L.R. and Li, Z., 2020. Identification of main-effect quantitative trait loci (QTLs) for low-temperature stress tolerance germination-and early seedling vigor-related traits in rice (Oryza sativa L.). Molecular Breeding, 40, pp.1-25.

Niu, Y. and Xiang, Y., 2018. An overview of biomembrane functions in plant responses to high-temperature stress. Frontiers in plant science, 9, p.915.

Oz, M.T., Eyidogan, F., Yucel, M. and Öktem, H.A., 2015. Functional role of nitric oxide under abiotic stress conditions. Nitric oxide action in abiotic stress responses in plants, pp.21-41.

Pang, J., Turner, N.C., Khan, T., Du, Y.L., Xiong, J.L., Colmer, T.D., Devilla, R., Stefanova, K. and Siddique, K.H., 2017. Response of chickpea (*Cicer arietinum* L.) to terminal drought: leaf stomatal conductance, pod abscisic acid concentration, and seed set. Journal of Experimental Botany, 68(8), pp.1973-1985.

Parankusam, S., Adimulam, S.S., Bhatnagar-Mathur, P. and Sharma, K.K., 2017. Nitric oxide (NO) in plant heat stress tolerance: current knowledge and perspectives. Frontiers in Plant Science, 8, p.1582.

Parmesan, C., Morecroft, M.D. and Trisurat, Y., 2022. Climate change 2022: Impacts, adaptation and vulnerability (Doctoral dissertation, GIEC).

Polsky, L. and von Keyserlingk, M.A., 2017. Invited review: Effects of heat stress on dairy cattle welfare. Journal of dairy science, 100(11), pp.8645-8657.

Pospíšil, P., 2016. Production of reactive oxygen species by photosystem II as a response to light and temperature stress. Frontiers in plant science, 7, p.1950.

Prasad, P. V. V., & Djanaguiraman, M. (2014). High night temperature decreases leaf photosynthesis and pollen function in grain sorghum. Functional Plant Biology, 41(12), 1269-1279.

Rai, K. K., Rai, N., & Rai, S. P. (2018). Salicylic acid and nitric oxide alleviate high temperature induced oxidative damage in *Lablab purpureus* L plants by regulating bio-physical processes and DNA methylation. Plant Physiology and Biochemistry, 128, 72-88.

Rai, K.K., Pandey, N. and Rai, S.P., 2020a. Salicylic acid and nitric oxide signaling in plant heat stress. Physiologia plantarum, 168(2), pp.241-255.

Rai, K.K., Rai, N., Aamir, M., Tripathi, D. and Rai, S.P., 2020. Interactive role of salicylic acid and nitric oxide on transcriptional reprogramming for high temperature tolerance in *Lablab purpureus* L.: Structural and functional insights using computational approaches. Journal of biotechnology, 309, pp.113-130.

Raza, A., Razzaq, A., Mehmood, S.S., Zou, X., Zhang, X., Lv, Y. and Xu, J., 2019. Impact of climate change on crops adaptation and strategies to tackle its outcome: A review. Plants, 8(2), p.34.

Rivero, R.M., Mittler, R., Blumwald, E. and Zandalinas, S.I., 2022. Developing climate-resilient crops: improving plant tolerance to stress combination. The Plant Journal, 109(2), pp.373-389.

Rykaczewska, K., 2015. The effect of high temperature occurring in subsequent stages of plant development on potato yield and tuber physiological defects. American Journal of Potato Research, 92, pp.339-349.

Schlenker, W., & Roberts, M. J. (2009). Nonlinear temperature effects indicate severe damages to US crop yields under climate change. Proceedings of the National Academy of Sciences, 106(37), 15594-15598.

Seki, M., Okamoto, M., Matsui, A., Kim, J.M., Kurihara, Y., Ishida, J., Morosawa, T., Kawashima, M., To, T.K. and Shinozaki, K., 2009. Microarray analysis for studying the abiotic stress responses in plants. Molecular Techniques in Crop Improvement: 2nd Edition, pp.333-355.

Setzer, J. and Higham, C., 2022. Global trends in climate change litigation: 2022 Snapshot.

Shekhawat, K., Almeida-Trapp, M., García-Ramírez, G.X. and Hirt, H., 2022. Beat the heat: Plant-and microbe-mediated strategies for crop thermotolerance. Trends in Plant Science.

Shinozaki, K. and Yamaguchi-Shinozaki, K., 2007. Gene networks involved in drought stress response and tolerance. Journal of experimental botany, 58(2), pp.221-227.

Singh, A., Mehta, S., Yadav, S., Nagar, G., Ghosh, R., Roy, A., Chakraborty, A. and Singh, I.K., 2022. How to cope with the challenges of environmental stresses in the era of global climate change: An update on ROS stave off in plants. International Journal of Molecular Sciences, 23(4), p.1995.

Singh, B., Kukreja, S. and Goutam, U., 2020. Impact of heat stress on potato (*Solanum tuberosum* L.): Present scenario and future opportunities. The Journal of Horticultural Science and Biotechnology, 95(4), pp.407-424.

Singh, K.M. and Singh, H.K., 2020. Effect of foliar application of potassium nitrate on late sown wheat (*Triticum aestivum* L.) in mitigating terminal heat stress. Journal of Pharmacognosy and Phytochemistry, 9(6S), pp.492-495.

Sun, Q., Miao, C., Hanel, M., Borthwick, A.G., Duan, Q., Ji, D. and Li, H., 2019. Global heat stress on health, wildfires, and agricultural crops under different levels of climate warming. Environment international, 128, pp.125-136.

Szymańska, R., Ślesak, I., Orzechowska, A. and Kruk, J., 2017. Physiological and biochemical responses to high light and temperature stress in plants. Environmental and Experimental Botany, 139, pp.165-177.

Tiwari, S., Lata, C., Singh Chauhan, P., Prasad, V. and Prasad, M., 2017. A functional genomic perspective on drought signaling and its crosstalk with phytohormone-mediated signaling pathways in plants. Current genomics, 18(6), pp.469-482.

Wahid, A., et al. (2007). Heat tolerance in plants: An overview. Environmental and Experimental Botany, 61(3), 199-223.

Wang, J., Song, L., Gong, X., Xu, J. and Li, M., 2020. Functions of jasmonic acid in plant regulation and response to abiotic stress. International Journal of Molecular Sciences, 21(4), p.1446.

Wen, J., Jiang, F., Weng, Y., Sun, M., Shi, X., Zhou, Y., Yu, L. and Wu, Z., 2019. Identification of heat-tolerance QTLs and high-temperature stress-responsive genes through conventional QTL mapping, QTL-seq and RNA-seq in tomato. BMC plant biology, 19, pp.1-17.

Wu, C., Tang, S., Li, G., Wang, S., Fahad, S. and Ding, Y., 2019. Roles of phytohormone changes in the grain yield of rice plants exposed to heat: a review. PeerJ, 7, p.e7792.

Yadav, A., Singh, J., Ranjan, K., Kumar, P., Khanna, S., Gupta, M., Kumar, V., Wani, S.H. and Sirohi, A., 2020. Heat shock proteins: master players for heat-stress tolerance in plants during climate change. Heat stress tolerance in plants: physiological, molecular and genetic perspectives, pp.189-211.

Yadav, M.R., Choudhary, M., Singh, J., Lal, M.K., Jha, P.K., Udawat, P., Gupta, N.K., Rajput, V.D., Garg, N.K., Maheshwari, C. and Hasan, M., 2022. Impacts, tolerance, adaptation, and mitigation of heat stress on wheat under changing climates. International Journal of Molecular Sciences, 23(5), p.2838.

Yang, J., Miao, W. and Chen, J., 2021. Roles of jasmonates and brassinosteroids in rice responses to high temperature stress–A review. The Crop Journal, 9(5), pp.977-985.

Yavuz, D., Seymen, M., Kal, Ü., Atakul, Z., Tanrıverdi, Ö.B., Türkmen, Ö. and Yavuz, N., 2023. Agronomic and physio-biochemical responses of lettuce to exogenous sodium nitroprusside (SNP) applied under different irrigation regimes. Agricultural Water Management, 277, p.108127.

Zandalinas, S.I., Fritschi, F.B. and Mittler, R., 2021. Global warming, climate change, and environmental pollution: recipe for a multifactorial stress combination disaster. Trends in Plant Science, 26(6), pp.588-599.

Zhang, J., Klueva, N.Y., Wang, Z., Wu, R., Ho, T.H.D. and Nguyen, H.T., 2000. Genetic engineering for abiotic stress resistance in crop plants. In Vitro Cellular & Developmental Biology-Plant, 36, pp.108-114.

Zhang, Z., Lan, M., Han, X., Wu, J. and Wang-Pruski, G., 2020. Response of ornamental pepper to high-temperature stress and role of exogenous salicylic acid in mitigating high temperature. Journal of Plant Growth Regulation, 39, pp.133-146.

# CHAPTER 2

# IMPORTANCE OF STUDYING HEAT STRESS IN PLANTS


## Abstract

With the increasing global temperatures and climate change, heat stress has become a significant concern for plant health and agriculture. Understanding the impact of heat stress on plants is crucial for developing strategies to mitigate its effects and ensure food security. Heat stress negatively affects crop productivity and can lead to reduced yields and economic losses for farmers. Scientists can identify the physiological and molecular mechanisms that contribute to plant heat tolerance by studying heat stress in plants. This knowledge can be used to develop heat-resistant crop varieties through breeding or genetic engineering approaches. By unravelling the genetic and physiological mechanisms involved in heat tolerance, scientists can identify potential targets for enhancing crop heat resistance. This knowledge can be utilised to develop improved crop management practices and breeding strategies that enable plants to thrive under high-temperature conditions.



**Keywords**: HT Stress, Antioxidants, Photoinhibition, Stomatal movement


## Introduction

Temperature above the optimal growing conditions that disrupt cellular homeostasis, thus severely affecting growth and the developmental process, is termed high temperature (HT) stress (Bita and Gerats, 2013). The rapidly increasing global temperature due to an increase in anthropogenic activity and natural means has become a serious threat to crop growth and productivity (Table 1) (Lobell et al., 2011). Being sessile, HT stress, along with other abiotic factors, potentially influence various metabolic processes in plants (Wahid et al., 2007). Nonetheless, plants execute a plethora of biological and molecular processes that regulate the expression of genes involved in the photosynthetic process and defense system, which substantially improve crop growth and yield (Fita et al., 2015). Several processes, such as the electron transport system, photorespiration, photosynthetic pigments, and $CO_2$ assimilatory pathway, are integral components of the photosynthetic apparatus, and failure of any one of them will hamper photosynthetic activity (Sharkey, 1988). Alteration of photosynthetic activity acts as a potent sensor of environmental perturbations to reflect changes in redox homeostasis and cellular energy imbalance (Allakhverdiev et al., 2008).

A plethora of studies has indicated that plants enhance their photosynthetic competency under high-temperature (HT) stress by adjusting their optimum growth temperature with seasonal temperature shifts (Berry and Bjorkman, 1980; Yamori et al., 2014). In addition, HT stress also significantly impacts seed germination, and the suppression of seed germination at HT is referred to as thermo-inhibition (Gómez et al., 2006). The germinated seedlings often show stunted growth and necrotic symptoms due to decreased photosynthetic activity upon exposure to high-temperature stress (Suzuki et al., 2012). HT stress also profoundly affects reproductive stages, severely impacting fertility. For example, a study reported a 50-70% decrease in spikelet fertility in rice plants exposed to 38°C, which otherwise showed 95% fertility during anthesis at the standard temperature of 29°C (Prasad et al., 2006). Similarly, another study reported a decrease in sugar phosphate levels during the grain development stage in rice plants growing under HT stress (Ruan et al., 2010). The decrease in sugar phosphate levels could have a negative impact on gluconeogenesis and the tricarboxylic acid (TCA) cycle, stimulating starch degradation and downregulating starch biosynthesis (Baud et al., 2009).

Specific membrane proteins and lipids are also quite sensitive to high-temperature (HT) stress (Mittler, 2006). The membranes are often composed of unsaturated fatty acids that can withstand HT stress, thereby increasing the tolerance of plants (Murata et al., 2007). Researchers have documented that altering fatty acids can stimulate plants' HT tolerance by modifying the membrane composition. In general, a strong membrane with improved fatty acid composition will enable enhanced fluidity, thus allowing proteins to move freely and provoking thermotolerance in plants (Larkindale and Huang, 2004). HT stress has been reported to alter the folding of proteins, stimulating their denaturation and thus preventing them from performing their regulatory functions (Vierling, 1991). Specific groups of proteins can regulate HT stress tolerance in plants by acting as molecular chaperones and enhancing protein scaffolding, popularly known as heat shock proteins (HSPs). HSPs are abundantly present in all living organisms. They are widely known to prevent HT stress-induced oxidative damage in plants by preventing protein denaturation and simultaneously stimulating the proper folding of proteins (Wang et al., 2004). This chapter comprehensively discusses the impact of HT stress on plants' physiological and biochemical processes. In addition, the protective measures adopted by plants to overcome HT stress-induced oxidative damage are also critically reviewed.

## HT stress-induced photoinhibition in plants

The HT stress, in conjunction with high light intensity, induces severe repercussions on the photosynthetic apparatus of plants by the process known as photoinhibition (Takahashi and Murata, 2008). Several studies have corroborated the above notion, where they have observed that light intensity also plays a crucial role in determining the severity of HT stress (Nouri et al., 2015; Yang et al., 2019). The temporal and spatial variation in light intensity, either by shading of clouds or by organisms or objects, modulates the growth and productivity of crop plants under HT stress (Nouri et al., 2015). Excess light intensity under HT conditions leads to a gradual decline in photosynthetic efficiency, which is mainly associated with the damaging of photosystem II (PSII), i.e., P680 (Takahashi and Murata, 2008). Being susceptible to high light intensity, HT stress differentially alters the rate of energy absorption and transfer, thus affecting the electron transport system (Yamori et al., 2014). The detailed mechanism by which high light intensity, accompanied by HT stress, negatively regulates photosynthesis is described below. Heat stress can impair the functioning of critical enzymes involved in photosynthesis, such as RuBisCO (Ribulose-1,5-bisphosphate carboxylase/oxygenase). As a result, the carbon dioxide fixation rate decreases, reducing photosynthetic efficiency. This reduction in efficiency can limit the plant's ability to produce sugars and other vital energy sources (Yamori et al., 2014). Recent reports have corroborated that high temperatures can lead to the production of excessive reactive oxygen species (ROS) within plant cells (Kollist et al., 2019). These ROS, including singlet oxygen, superoxide radicals, and hydrogen peroxide, can cause oxidative damage to cellular components, including chlorophyll molecules, proteins, lipids, and DNA. Such damage impairs the photosynthetic apparatus and disrupts the light-capturing and energy conversion processes (Kollist et al., 2019).

**Table 1** Importance of studying heat stress in plants

| Importance | Description | References |
|---|---|---|
| Understanding Climate Change Impact | Studying heat stress in plants helps us comprehend the effects of climate change on agricultural systems and develop strategies for adaptation. | IPCC. (2014). |
| Ensuring Global Food Security | Heat stress can lead to reduced crop yields and quality, highlighting the need to understand its impact and develop heat-tolerant crop varieties. | Lobell and Gourdji (2012) |
| Developing Heat-Tolerant Crop Varieties | By studying heat stress responses in plants, we can identify genetic traits and mechanisms that contribute to heat tolerance and breed resilient crop varieties. | Prasad and Djanaguiraman (2011) |
| Enhancing Crop Resilience | Understanding the physiological and molecular responses of plants to heat stress helps develop strategies to enhance their resilience and mitigate yield losses. | Hasanuzzaman et al. (2013) |
| Improving Crop Management Practices | Knowledge of heat stress in plants allows for the development of improved management practices, including irrigation, shade, and mulching, to mitigate its impact. | Hatfield et al. (2011) |
| Unraveling Molecular Mechanisms of Heat Stress | Studying heat stress helps unravel the molecular mechanisms underlying plant responses, such as gene expression, heat shock proteins, and antioxidant defense. | Haider et al. (2022). |

## Effect on photosynthetic apparatus

Under normal conditions, the rate of energy absorption and transfer across the photosystem and electron transport is highly stabilized, fulfilling fundamental thermodynamic requirements (Yang et al., 2019). The slow and stable energy transfer minimizes the uphill reactions, which are then exploited in the electron transport system for ATP synthesis in the form of NADPH (Kollist et al., 2019). Photoinhibition in plants occurs mainly when the level of ATP synthesis and other reducing powers fails to meet the demand of the carbon reduction cycle. In addition, the level of NADP+ also declines sharply, preventing it from being an effective electron acceptor of linear photosynthetic systems (Kollist et al., 2019). Under HT stress conditions coupled with high light intensity, the photosynthetic reaction centers, specifically PSII and PSI, become saturated, causing an imbalance in the fraction of energy used and unused energy. The new energy, technically referred to as excitation energy, is potentially harmful to the photosynthetic membrane (Yang et al., 2019). The excitation energy may cause the oxidation of essential pigments and amino acids, leading to the degradation of D1 proteins (Takahashi and Murata, 2008).

Furthermore, the excitation energy-induced oxidation of plastoquinone disrupts electron flow from P680 to plastoquinone, resulting in the formation of P680 triplet, which upon reacting with oxygen, generates singlet oxygen, a free radical with devastating properties (Pospíšil, 2009). The generated singlet oxygen will, in turn, oxidize P680, leading to a progressive decrease in the number of active PSII units, inevitably causing photoinhibition (Takahashi and Murata, 2008). In addition, the bleaching of P680 is also associated with the decline of the electron transport chain as it fails to repair the damaged D1 proteins (Takahashi and Murata, 2008). High temperatures can lead to the inactivation or damage of the PSII reaction centers, particularly the D1 protein subunit. The D1 protein is crucial for the efficient transfer of electrons in PSII. When it becomes damaged or denatured due to heat stress, the electron transfer process is impaired, reducing the efficiency of PSII (Allakhverdiev et al., 2008). Several studies have also confirmed that HT stress can lead to the disruption of the oxygen-evolving complex (OEC) (Yamamoto, 2016). The OEC in PSII is responsible for splitting water molecules and releasing oxygen during photosynthesis. Heat stress can disrupt the stability and function of the OEC, resulting in reduced water-splitting activity. As a result, electron flow and ATP production are compromised (Yamamoto, 2016).

Similarly, PSI plays a crucial role in accepting electrons from PSII and transferring them to the final electron acceptor, ultimately leading to the production of NADPH (Shikanai, 2007). HT stress can disrupt the electron transfer process between PSII and PSI. The electron flow from PSII to PSI can become unbalanced, leading to an accumulation of electrons in the electron transport chain (Baker and Oxborough, 2004). This imbalance can cause an overproduction of the electron carriers and the formation of reactive oxygen species, damaging PSI components and inhibiting its activity (Shikanai, 2007). Heat stress can destabilize the protein complexes involved in PSI assembly and function (Sonoike, 2011). This can reduce functional PSI complexes, limiting the plant's ability to transfer electrons and generate NADPH. Furthermore, PSI is also involved in cyclic electron flow, which causes additional ATP molecules beyond what is produced by linear electron flow (Shikanai, 2007). Heat stress can disrupt the cyclic electron flow and reduce the production of ATP, further impacting the energy availability for various metabolic processes (Yamori et al., 2014).

## Impact on the electron transport chain

HT stress can significantly impact plants' electron transport chain (ETC). The ETC is a critical component of photosynthesis, the process by which plants convert light energy into chemical energy. It takes place in the thylakoid membranes of chloroplasts, and heat stress can disrupt various steps of this process, leading to adverse effects on plant growth and productivity (Allakhverdiev et al., 2008). HT stress can affect the expression of genes that encode critical proteins involved in the ETC. HT can alter gene expression patterns, leading to changes in the abundance and activity of ETC-related proteins (Pospíšil, 2009). This disruption can impact the assembly and function of ETC complexes, such as photosystem II (PSII), photosystem I (PSI), cytochrome b6f complex, and ATP synthase (Shikanai, 2007).

A large body of literature has shown that HT induces the expression of heat shock proteins, which play a crucial role in protecting cellular components, including ETC proteins, from thermal damage (Sonoike, 2011). Heat shock proteins can help stabilize and refold denatured proteins, prevent protein aggregation, and maintain proper protein structure and function under heat-stress conditions. By facilitating the appropriate folding and assembly of ETC proteins, heat shock proteins can mitigate the detrimental effects of heat stress on the electron transport chain (Yamamoto, 2016).

The ETC is crucial in generating ATP, the cell's energy currency. Heat stress can disrupt ATP synthesis by affecting the proton gradient across the thylakoid membrane (Hu et al., 2020). High temperatures can alter the activity of ATP synthase, the enzyme responsible for ATP synthesis, leading to reduced ATP production. This can limit the energy availability for various metabolic processes in plants (Bita and Gerats, 2013). Heat stress can exacerbate photoinhibition, a process in which excessive light energy damages the photosynthetic apparatus (Wahid et al., 2007). When plants experience HT, the balance between light absorption and energy utilization can be disrupted. This can lead to the overexcitation of the photosystems, increased production of ROS, and ultimately, damage to the ETC components (Allakhverdiev et al., 2008). HT stress can disrupt the integral components of the Calvin cycle, which is responsible for fixing carbon dioxide and producing carbohydrates (Kollist et al., 2019). The Calvin cycle exploits the products of the ETC, such as ATP and NADPH, to catalyze the synthesis of sugars from $CO_2$ (Foyer and Noctor, 2009). Heat stress-induced disruptions in the ETC can lead to reduced availability of these energy-rich molecules, thereby limiting carbon fixation and overall plant growth (Allakhverdiev et al., 2008).

HT stress can also modulate post-translational modifications of regulatory proteins by stimulating protein denaturation and unfolding (Chen et al., 2017). To restore protein homeostasis, plants upregulate the expression of molecular chaperones, such as HSP70 and HSP90, which assist in proper protein folding, assembly, and transport (Tracz-Gaszewska et al., 2017). At the same time, the molecular chaperones regulate the expression of another class of proteins called proteases, which function as host clean-up, i.e., are activated to degrade misfolded or damaged proteins (Per et al., 2018). These chaperones and proteases play a critical role in maintaining the integrity and functionality of ETC proteins under heat stress conditions (Tracz-Gaszewska et al., 2017). Overall, heat stress can disrupt essential genes and proteins involved in the electron transport chain in plants. It can alter gene expression patterns, affect the abundance and activity of ETC-related proteins, induce the expression of heat shock proteins and ROS scavenging enzymes, and activate chaperones and proteases to mitigate protein damage. These cellular responses aim to maintain the electron transport chain's functionality and minimize heat stress's detrimental effects on plant metabolism and growth.

## Effect on stomatal opening

HT stress can indeed affect stomatal movement in plants. Stomata are small openings or pores primarily found on the surfaces of plant leaves that regulate gas exchange, allowing for the uptake of carbon dioxide ($CO_2$) and the release of oxygen ($O_2$) during photosynthesis (Hetherington and Woodward, 2003). They also play a crucial role in regulating water loss from the plant through transpiration (Noctor et al., 2012). Heat stress often leads to stomata's closure, reducing the gas exchange rate. This closure occurs as a protective mechanism to prevent excessive water loss through transpiration (González-Schain et al., 2016). By closing the stomata, plants can conserve water and reduce the risk of dehydration (Lawson and Blatt, 2014). Additionally, stomatal closure due to HT stress restricts the movement of carbon dioxide into the leaf, thereby negatively impacting the rate of photosynthesis (Yamori et al., 2014). Reduced photosynthesis can adversely affect plant growth and productivity (Bita and Gerats, 2013).

Heat stress can lead to a dramatic increase in the surface temperature of a leaf. Higher leaf surface temperatures can further exacerbate stomatal closure as a response to HT stress (Raschke, 1975). This closure can lead to an accumulation of internal carbon dioxide, thus restricting the ability of photosynthesis (Lawson and Blatt, 2014). In addition, HT stress can disrupt the balance of plant hormones involved in stomatal regulation (Per et al., 2018). For example, abscisic acid (ABA), a stress hormone, is often synthesized and accumulated in response to heat stress. ABA promotes stomatal closure and helps plants cope with heat stress. However, if the ABA levels become excessively high, it can lead to prolonged stomatal closure and negatively impact photosynthesis (Alsufyani et al., 2020). Under HT stress conditions, abscisic acid (ABA) plays a crucial role in regulating stomatal movements in plants. ABA is a plant hormone that is involved in various physiological processes, including stomatal regulation, especially under stressful conditions (Christmann et al., 2013). ABA accumulation is stimulated under heat stress is well known in plants. HT triggers the synthesis and release of ABA within the plant. Elevated levels of ABA promote stomatal closure, reducing the aperture size of the stomatal pore. Stomatal closure helps plants conserve water and limit transpiration, thus preventing excessive water loss under heat stress (Alsufyani et al., 2020)

Correspondingly, ABA also modulates various signaling pathways in guard cells, cells that surround stomata and control their opening and closing. Studies have corroborated that ABA-induced stomatal closure is initiated after binding of ABA with specific receptors in the plasma membrane of guard cells (Verslues and Zhu, 2007). This signaling cascade involves modulation of ion homeostasis, particularly the efflux of potassium ions ($K^+$) and anions, which results in osmotic water loss from the guard cells, leading to their shrinking and subsequent stomatal closure (Higashi et al., 2021). ABA also indirectly affects stomatal movements under heat stress by regulating other factors involved in stomatal regulation. For example, ABA can regulate the expression and activity of ion channels, such as potassium channels and anion channels, which are critical for controlling guard cell turgor and stomatal aperture (Nag et al., 2009). ABA-mediated modulation of these channels can induce inevitable alterations in ion fluxes and subsequent changes in guard cell volume and stomatal aperture, thus affecting the overall stability and processivity of ETC and photosynthesis (Merilo et al., 2015). ABA can also modulate stomatal movement after interacting with other phytohormones, such as salicylic acid, jasmonic acid, nitric oxide, and ethylene, to coordinate stomatal responses to heat stress (Merilo et al., 2015). These hormones may act synergistically or antagonistically with ABA to regulate stomatal movements. The exact interactions and cross-talk mechanisms can vary depending on the plant species and specific environmental conditions. In response to heat stress, ABA accumulation promotes stomatal closure through several physiological and biochemical mechanisms. This closure helps plants conserve water, maintain cellular hydration, and cope with heat stress conditions. Understanding the role of ABA in stomatal regulation is crucial for developing strategies to improve plant heat tolerance and mitigate the adverse effects of heat stress on crop productivity.

## HT stress-mediated generation of reactive oxygen species (ROS)

Reactive oxygen species (ROS) play essential roles in plant biology. They are chemically reactive molecules containing oxygen, such as superoxide radicals ($O2^{\bullet -}$), hydrogen peroxide ($H_2O_2$), and hydroxyl radicals ($^{\bullet}OH$). ROS are produced as natural byproducts of various metabolic processes in plants, including photosynthesis, respiration, and enzymatic reactions (Mittler, 2017). Exaggerated synthesis of ROS is potentially damaging to cells due to their high reactivity; plants have evolved intricate mechanisms to control and utilize ROS for their benefit (Mignolet-Spruyt et al., 2016). When their levels are below a certain threshold, they function as signaling molecules involved in various physiological processes, including growth, development, and responses to environmental stress (Wei et al., 2015). Recent findings have shown that ROS act as signaling molecules in plant cells, regulating various cellular and physiological processes such as seed germination, stomatal closure, programmed cell death, and defense responses (Suzuki et al., 2012). They can trigger specific signaling pathways, including activation of protein kinases and transcription factors, leading to enhanced expression of defense-responsive genes/proteins (Considine and Foyer, 2014).

Studies have indicated that ROS production is crucial to a plant's defense response against biotic/abiotic stresses. When plants detect stress anomalies, they generate ROS at the target site to induce programmed cell death, known as the hypersensitive response. This localized cell death restricts the biotrophic attack (Bawa et al., 2022). ROS are also known to intricately regulate plant responses to various abiotic stresses, including high light, drought, salinity, and heat stress (Gill and Tuteja, 2010). Upon exposure to abiotic stress, ROS are produced, and at the basal level, they act as signaling molecules, thus initiating stress signaling cascades (Suzuki et al., 2012). ROS can activate stress-responsive genes and antioxidant defense systems, enhancing the plant's ability to tolerate and adapt to adverse conditions (Table 2).

**Table 2** Role of enzymatic and non-enzymatic antioxidants in improving heat stress tolerance in plants

| Antioxidant | Role | Protective Mechanisms | References |
|---|---|---|---|
| **Enzymatic** | | | |
| Superoxide Dismutase (SOD) | Converts superoxide radicals into hydrogen peroxide, preventing oxidative damage. | Scavenging of superoxide radicals, maintenance of cellular redox balance, activation of other antioxidant enzymes. | Mittler (2002) |
| Catalase (CAT) | Detoxifies hydrogen peroxide by converting it into water and oxygen. | Prevention of hydrogen peroxide accumulation, protection against oxidative stress-induced damage. | Mhamdi et al. (2010) |
| Ascorbate Peroxidase (APX) | Converts hydrogen peroxide into water using ascorbate as a reducing agent. | Maintenance of ascorbate redox state, protection against oxidative damage, regeneration of other antioxidants such as glutathione. | Foyer and Noctor (2011) |
| Glutathione Reductase (GR) | Maintains the reduced form of glutathione (GSH) by regenerating it from its oxidized form (GSSG). | Recycling of GSH for continued antioxidant activity, protection against oxidative stress. | Noctor et al. (2012) |
| Glutathione Peroxidase (GPX) | Catalyzes the reduction of hydrogen peroxide and organic hydroperoxides using GSH as a reducing agent. | Detoxification of peroxides, protection of cellular components, maintenance of redox homeostasis. | Hasanuzzaman et al. (2013) |

| | | | |
|---|---|---|---|
| Dehydroascorbate Reductase (DHAR) | Converts dehydroascorbate (DHA) back to ascorbate (AsA), ensuring the regeneration of AsA. | Maintenance of AsA pool, protection against oxidative stress, regeneration of other antioxidants. | Pignocchi et al. (2006) |
| Monodehydroascorbate Reductase (MDHAR) | Reduces monodehydroascorbate (MDHA) to AsA, preventing the accumulation of harmful reactive intermediates. | Regeneration of AsA pool, prevention of oxidative damage, maintenance of redox homeostasis. | Foyer et al. (1998) |
| **Non-enzymatic** | | | |
| Ascorbic Acid (AsA) | Acts as a direct scavenger of reactive oxygen species and regenerates other antioxidants. | Reduction of oxidative stress, maintenance of redox balance, protection of cellular components. | Hasanuzzaman et al. (2013) |
| Glutathione (GSH) | Participates in various redox reactions and acts as a substrate for glutathione peroxidases. | Regulation of cellular redox status, detoxification of reactive oxygen species, protection against oxidative damage. | Noctor et al. (2012). |
| α-Tocopherol (Vitamin E) | Scavenges lipid peroxyl radicals in membranes and protects against lipid peroxidation. | Protection of cell membranes and chloroplasts from oxidative damage, maintenance of membrane fluidity. | Mène-Saffrané and DellaPenna (2010). |
| Proline (Pro) | Acts as a compatible solute, stabilizes proteins and membranes, and scavenges reactive oxygen species. | Osmotic adjustment, protection of protein structure, scavenging of hydroxyl radicals and singlet oxygen. | Ashraf and Foolad (2007) |

## Free radicals

Free radicals in plants refer to highly reactive molecules that contain unpaired electrons in their outermost shell. These molecules are generated as natural byproducts of various metabolic processes in plants, including photosynthesis, respiration, and enzymatic reactions (Foyer and Noctor, 2009; Apel and Hirt, 2004). Free radicals can also be formed in plants as a response to environmental stresses such as UV radiation, pollutants, and pathogen attacks. Common types of free radicals in plants include superoxide radicals ($O2^{\bullet-}$) and hydroxyl radicals ($^{\bullet}OH$) (Chiou and Lin, 2011; Apel and Hirt, 2004). These free radicals can cause damage to plant cells and tissues through a process known as oxidative stress. They react with cellular components, including lipids, proteins, and DNA, leading to cell membrane disruption, enzyme inactivation, and genetic mutations (Gill and Tuteja, 2010; Sharma et al., 2012).

However, plants have evolved various defense mechanisms to counteract the harmful effects of free radicals and maintain cellular redox balance (Noctor et al., 2016). One of the primary defense systems is the production of antioxidants, which scavenge and neutralize free radicals. Antioxidants in plants include enzymes like superoxide dismutase (SOD), catalase (CAT), and peroxidase (POD), as well as non-enzymatic compounds such as ascorbic acid (vitamin C), tocopherols (vitamin E), carotenoids, and flavonoids (Apel and Hirt, 2004; Gill and Tuteja, 2010).

These antioxidants work in a coordinated manner to prevent the accumulation of free radicals and minimize oxidative damage (Chiou and Lin, 2011). They donate electrons or hydrogen atoms to free radicals, stabilizing them and breaking the chain reaction of oxidation. Additionally, plants have mechanisms to repair or eliminate damaged molecules caused by free radicals, ensuring proper cell function and overall plant health (Gill and Tuteja, 2010). While free radicals can have detrimental effects on plants, their generation is a natural part of plant metabolism. Through antioxidant defense systems, plants maintain a delicate balance between free radical generation and neutralization, allowing them to survive and thrive in various environmental conditions.

## Non-radicals

Hydrogen peroxide ($H_2O_2$) and singlet oxygen ($^1O_2$) are the two non-radicals that play essential roles in plant biology, both as signaling molecules and as contributors to oxidative stress (Castro et al., 2021;

Triantaphylidès and Havaux, 2009). Hydrogen peroxide is a relatively stable and diffusible reactive oxygen species (ROS) in plants. It is produced in various cellular compartments, including chloroplasts, peroxisomes, and mitochondria. $H_2O_2$ acts as a signaling molecule in several physiological processes, such as seed germination, stomatal closure, and defense responses (Rai et al., 2019; Sharma et al., 2012). H2O2 regulates gene expression, protein phosphorylation, and cellular processes related to plant growth, development, and stress responses. It can activate specific signaling pathways, leading to the expression of genes involved in stress tolerance and defense mechanisms (Castro et al., 2021; Sharma et al., 2012). $H_2O_2$ also activates plant hormone signaling, including auxin, abscisic acid (ABA), and salicylic acid (SA). When produced excessively or under certain stress conditions, $H_2O_2$ can contribute to oxidative stress (Suzuki et al., 2012). It can react with cellular components, such as lipids, proteins, and DNA, causing damage and disrupting cellular functions. However, plants have antioxidant systems, including enzymes (e.g., catalase, peroxidase) and non-enzymatic compounds (e.g., ascorbic acid, glutathione), that help scavenge and detoxify excess $H_2O_2$, maintaining cellular redox balance (Apel and Hirt, 2004; Gill and Tuteja, 2010).

Singlet oxygen is a highly reactive and short-lived form of oxygen ($^1O_2$). It is generated during photosynthesis, particularly under high-light conditions or when the photosynthetic machinery is stressed. Singlet oxygen can cause oxidative damage to cellular components and is associated with photoinhibition and photooxidative stress (Ramel et al., 2019). Singlet oxygen is primarily produced in the chloroplasts' photosystem II (PSII) during photosynthetic electron transport. Excess energy or electron leakage within PSII can result in the formation of singlet oxygen, which can harm the photosynthetic apparatus and lead to the degradation of chlorophyll and other cellular components (Mittler 2017).

To mitigate ROS-induced oxidative damages, plants have evolved stress defense mechanisms. One defense mechanism involves scavenging singlet oxygen by carotenoids, which act as quenchers and dissipate excess energy. Carotenoids, such as β-carotene and lutein, are efficient singlet oxygen quenchers and are present in the photosynthetic apparatus to protect against singlet oxygen-induced oxidative stress (Ramel et al., 2019). In summary, both free radicals and non-radicals are reactive oxygen species generated in plants, serving as essential signaling molecules but also potentially contributing to oxidative stress. Plants have developed antioxidant systems and defense mechanisms to regulate and neutralize these reactive species, maintaining cellular redox homeostasis and minimizing oxidative damage.

## Strategies adopted by plants to ameliorate HT-induced oxidative damages

Plants have developed several strategies to mitigate the oxidative damages caused by HT stress. These adaptive mechanisms help maintain cellular homeostasis, protect essential cellular components, and minimize the detrimental effects of oxidative stress (Mittler, 2017). Here are some strategies adopted by plants to ameliorate HT-induced oxidative damage (Figure 1).

(a) Activation of antioxidant defense system: Plants enhance the activity and expression of enzymatic antioxidants such as superoxide dismutase (SOD), catalase (CAT), ascorbate peroxidase (APX), glutathione peroxidase (GPX), and glutathione reductase (GR). Increased activity of these enzymes helps to scavenge ROS and prevent oxidative damage (Gill and Tuteja, 2010).

(b) Accumulation of compatible solutes: Plants accumulate compatible solutes or osmoprotectants such as proline, betaine, and sugars in response to HT stress. These solutes act as antioxidants and help stabilize cellular structures, maintain water balance, and protect proteins and membranes from oxidative damage (Ashraf and Harris, 2013).

(c) Heat shock proteins (HSPs): Heat shock proteins are chaperones that assist in protein folding, repair damaged proteins, and prevent protein aggregation under HT stress. They play a crucial role in protecting proteins from heat-induced denaturation and oxidative damage (Wang et al., 2004).

(d) Activation of ROS scavengers: Plants activate non-enzymatic antioxidants, including ascorbic acid (vitamin C), tocopherols (vitamin E), carotenoids, and phenolic compounds. These antioxidants directly scavenge ROS and protect cellular components from oxidative stress (Wei et al., 2015).

(e) Enhanced lipid metabolism: HT stress can cause lipid peroxidation, resulting in the production of reactive aldehydes and peroxides. Plants respond by upregulating lipid metabolism enzymes such as lipoxygenase (LOX) and phospholipase D (PLD), which help repair damaged membranes and prevent further lipid peroxidation (Ashraf and Harris, 2013).

(f) Regulation of ROS production: Plants regulate ROS production by modulating the activity of enzymes involved in electron transport chains, such as NADPH oxidases (respiratory burst oxidase homologs, RBOHs). Controlling ROS generation helps maintain redox homeostasis and prevents excessive oxidative damage (Foyer and Noctor, 2016).

(g) Activation of signaling pathways: Plants activate various signaling pathways, including mitogen-activated protein kinases (MAPKs), calcium-mediated signaling, and stress-responsive transcription factors (e.g., heat shock factors, HSFs), which coordinate the expression of stress-responsive genes involved in antioxidant defense and stress tolerance (Suzuki et al., 2012).

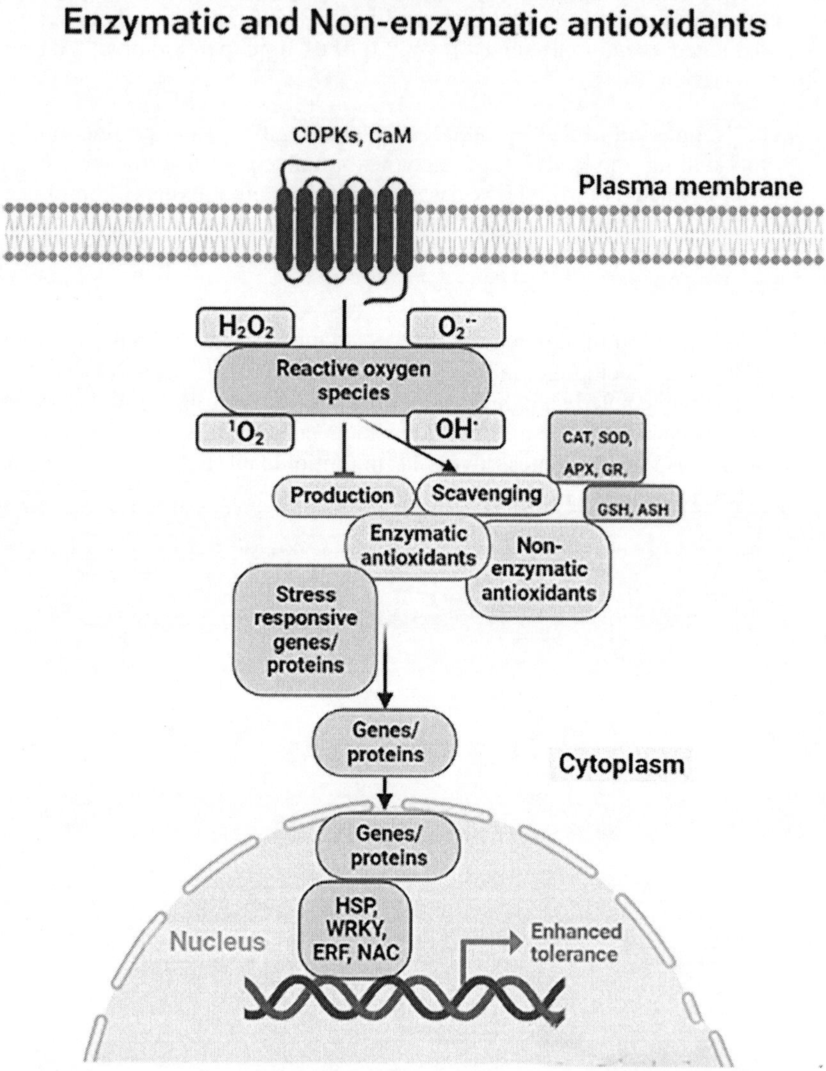


**Figure 1** An overview of enzymatic and non-enzymatic mediated scavenging of reactive oxygen species in plants exposed to HT stress. $H_2O_2$, hydrogen peroxide; $O_2^{\cdot-}$, superoxide anion; $OH^{\cdot}$, hydroxyl radical; $^1O_2$, singlet oxygen; CAT: Catalase; SOD: Superoxide dismutase; GR, Glutathione reductase; APX, Ascorbate peroxidase; GSH, Reduced glutathione; ASH, Reduced ascorbate; HSP, Heat shock protein; ERF, Ethylene responsive factor; NAC, No Apical Meristem, Arabidopdis Transcription Activator Factor, Cup Shaped Cotyledon

## Enzymatic antioxidants

Plants possess a variety of enzymatic antioxidants that play crucial roles in protecting them against oxidative stress. These enzymatic antioxidants are part of the plant's defence system and help maintain cellular homeostasis by scavenging harmful ROS. Here are some of the critical enzymatic antioxidants found in plants:

(a) Superoxide dismutase (SOD): SOD is an enzyme that catalyzes the conversion of superoxide radicals ($O_2^-$) into hydrogen peroxide ($H_2O_2$) and molecular oxygen ($O_2$). It acts as the first line of defense against ROS by neutralizing the highly reactive superoxide radicals (Mittler, 2017).

(b) Catalase (CAT): Catalase is an enzyme that decomposes hydrogen peroxide ($H_2O_2$) into water ($H_2O$) and molecular oxygen ($O_2$). It helps detoxify hydrogen peroxide, which is another harmful ROS, preventing the formation of more reactive species like hydroxyl radicals (Mhamdi et al., 2010).

(c) Ascorbate peroxidase (APX): APX is an enzyme that uses ascorbic acid (vitamin C) as a cofactor to break down hydrogen peroxide ($H_2O_2$) into water ($H_2O$) and monodehydroascorbate (MDHA). It plays a significant role in the ascorbate-glutathione cycle, which helps maintain the redox balance in plant cells (Pandey et al., 2017).

(d) Glutathione peroxidase (GPX): GPX is an enzyme that utilizes glutathione (GSH) as a reducing agent to detoxify hydrogen peroxide ($H_2O_2$) and organic hydroperoxides. It converts them into water ($H_2O$) and corresponding alcohols, protecting cells from oxidative damage (Foyer and Noctor, 2016).

(e) Glutathione reductase (GR): GR is an enzyme that maintains the pool of reduced glutathione (GSH) by catalyzing the conversion of oxidized glutathione (GSSG) back into its reduced form (GSH). GSH is an essential antioxidant in plants and participates in various redox reactions (Noctor et al., 2016). These enzymatic antioxidants work together to scavenge and neutralise reactive oxygen species, which are generated as byproducts of normal plant metabolism or in response to various stress conditions. By doing so, they help prevent oxidative damage to cellular components and maintain plants' overall health and vitality.

## Non-enzymatic antioxidants

Non-enzymatic antioxidants in plants are a diverse group of compounds that protect plants against oxidative stress and damage caused by ROS. These antioxidants are not directly involved in enzymatic reactions but are scavengers of free radicals and other harmful molecules. Here are some essential non-enzymatic antioxidants found in plants:

(a) Phenolic Compounds: This is a large and diverse group of antioxidants, including flavonoids (such as quercetin, kaempferol, and catechins), phenolic acids (such as ferulic acid and gallic acid), and tannins. Phenolic compounds are abundant in fruits, vegetables, and other plant-based foods (Pandey and Rizvi, 2009).

(b) Carotenoids: Carotenoids are pigments responsible for many fruits and vegetables' red, orange, and yellow colors. They include beta-carotene, lycopene, lutein, and zeaxanthin. Carotenoids act as antioxidants by quenching singlet oxygen and scavenging other ROS (Fiedor and Burda, 2014).

(c) Vitamin C (Ascorbic Acid): Ascorbic acid is a water-soluble and potent antioxidant found in many fruits and vegetables. It can donate electrons to neutralize free radicals and regenerate other antioxidants, such as vitamin E (Andre et al., 2010).

(d) Vitamin E (Tocopherols and Tocotrienols): Vitamin E is a group of fat-soluble antioxidants that protect cell membranes from oxidative damage. The primary forms of vitamin E in plants are alpha-tocopherol and gamma-tocopherol (Ahmad et al., 2022).

(e) Glutathione: Glutathione is a tripeptide composed of three amino acids: cysteine, glycine, and glutamic acid. It acts as an essential cellular antioxidant by directly scavenging ROS and participating in enzymatic reactions that regenerate other antioxidants (Noctor et al., 2016).

(f) Alkaloids: Some alkaloids, such as caffeine, theobromine, and nicotine, have antioxidant properties. These compounds are commonly found in plants and often serve as natural defense mechanisms against herbivores (Noctor et al., 2012).

(g) Phytoalexins: Phytoalexins are antimicrobial compounds produced by plants in response to pathogen attacks or other stressors. They often possess antioxidant activity and contribute to the overall defense system of plants (Dixon, 2001). These non-enzymatic antioxidants work together to scavenge and neutralise ROS, protecting plants from oxidative damage and maintaining their physiological functions. These compounds have also been linked to potential health benefits for humans when consumed through plant-based foods.

## Conclusion

Plants are vital in maintaining the ecological balance and sustaining the environment. Heat stress can disrupt the functioning of plants, including their ability to carry out photosynthesis, nutrient uptake, and water regulation. Understanding the impact of heat stress on plant ecosystems is crucial for preserving biodiversity and the environment's overall health. Heat stress in plants can have indirect effects on human health. For example, heat stress-induced reductions in crop yields can lead to food shortages and increased food prices, which can impact the nutrition and well-being of populations. Understanding the effects of heat stress on plants helps us to develop strategies to ensure a stable food supply and minimise the potential negative impacts on human health. Plants form the foundation of terrestrial ecosystems, and their response to heat stress influences ecosystems' overall resilience and stability. By studying how different plant species and communities respond to heat stress, scientists can assess the potential impacts on ecosystem dynamics, such as changes in species composition, nutrient cycling, and carbon sequestration. Overall, studying heat stress in plants is crucial for understanding the impacts of climate change, maintaining crop productivity, preserving environmental sustainability, developing adaptation strategies, safeguarding human health, and promoting ecosystem resilience. The knowledge gained from such studies can help devise effective strategies to mitigate the detrimental effects of heat stress on plants and ensure a sustainable and resilient future.

## References

Ahmad, A., Blasco, B. and Martos, V., 2022. Combating salinity through natural plant extracts based biostimulants: a review. Frontiers in Plant Science, 13, p.862034.

Allakhverdiev, S. I., Kreslavski, V. D., Klimov, V. V., Los, D. A., Carpentier, R., and Mohanty, P. (2008). Heat stress: an overview of molecular responses in photosynthesis. Photosynthesis Research, 98(1-3), 541-550.

Allakhverdiev, S.I., Kreslavski, V.D., Klimov, V.V., Los, D.A., Carpentier, R. and Mohanty, P., 2008. Heat stress: an overview of molecular responses in photosynthesis. Photosynthesis research, 98, pp.541-550.

Alsufyani, T., Califano, G., Deicke, M., Grueneberg, J., Weiss, A., Engelen, A.H., Kwantes, M., Mohr, J.F., Ulrich, J.F. and Wichard, T., 2020. Macroalgal–bacterial interactions: identification and role of thallusin in morphogenesis of the seaweed Ulva (Chlorophyta). Journal of experimental botany, 71(11), pp.3340-3349.

Andre, C.M., Larondelle, Y. and Evers, D., 2010. Dietary antioxidants and oxidative stress from a human and plant perspective: a review. Current Nutrition & Food Science, 6(1), pp.2-12.

Apel, K. and Hirt, H., 2004. Reactive oxygen species: metabolism, oxidative stress, and signal transduction. Annu. Rev. Plant Biol., 55, pp.373-399.

Ashraf, M., & Foolad, M. R. (2007). Roles of glycine betaine and proline in improving plant abiotic stress resistance. Environmental and Experimental Botany, 59(2), 206-216.

Ashraf, M., & Harris, P. J. (2013). Photosynthesis under stressful environments: An overview. Photosynthetica, 51(2), 163-190.

Baker, N.R. and Oxborough, K., 2004. Chlorophyll fluorescence as a probe of photosynthetic productivity. In Chlorophyll a fluorescence: a signature of photosynthesis (pp. 65-82). Dordrecht: Springer Netherlands.

Baud, S., Wuilleme, S., To, A., Rochat, C. and Lepiniec, L., 2009. Role of WRINKLED1 in the transcriptional regulation of glycolytic and fatty acid biosynthetic genes in Arabidopsis. The Plant Journal, 60(6), pp.933-947.

Bawa, G., Liu, Z., Zhou, Y., Fan, S., Ma, Q., Tissue, D.T. and Sun, X., 2022. Cotton proteomics: Dissecting the stress response mechanisms in cotton. Frontiers in Plant Science, 13, p.1035801.

Berry, J., & Bjorkman, O. (1980). Photosynthetic response and adaptation to temperature in higher plants. Annual Review of Plant Physiology, 31(1), 491-543.

Bita, C. E., & Gerats, T. (2013). Plant tolerance to high temperature in a changing environment: scientific fundamentals and production of heat stress-tolerant crops. Frontiers in Plant Science, 4, 273. doi: 10.3389/fpls.2013.00273

Castro, B., Citterico, M., Kimura, S., Stevens, D.M., Wrzaczek, M. and Coaker, G., 2021. Stress-induced reactive oxygen species compartmentalization, perception and signaling. Nature plants, 7(4), pp.403-412.

Chen, J., Lin, B., Huang, Q., Hu, L., Zhuo, K. and Liao, J., 2017. A novel Meloidogyne graminicola effector, MgGPP, is secreted into host cells and undergoes glycosylation in concert with proteolysis to suppress plant defenses and promote parasitism. PLoS pathogens, 13(4), p.e1006301.

Chiou, T.J. and Lin, S.I., 2011. Signaling network in sensing phosphate availability in plants. Annual review of plant biology, 62, pp.185-206.

Christmann, A., Weiler, E. W., Steudle, E., & Grill, E. (2007). A hydraulic signal in root-to-shoot signaling of water shortage. The Plant Journal, 52(1), 167-174. doi:

Considine, M. J., & Foyer, C. H. (2014). Redox regulation of plant development. Antioxidants & Redox Signaling, 21(9), 1305-1336. doi: 10.1089/ars.2013.5665

Dixon, R. A. (2001). Natural products and plant disease resistance. Nature, 411(6839), 843-847.

Fiedor, J., & Burda, K. (2014). Potential role of carotenoids as antioxidants in human health and disease. Nutrients, 6(2), 466-488.

Fita, A., Rodríguez-Burruezo, A., Boscaiu, M., Prohens, J. and Vicente, O., 2015. Breeding and domesticating crops adapted to drought and salinity: a new paradigm for increasing food production. Frontiers in Plant Science, 6, p.978.

Foyer, C. H., & Noctor, G. (2009). Redox regulation in photosynthetic organisms: signaling, acclimation, and practical implications. Antioxidants & Redox Signaling, 11(4), 861-905.

Foyer, C. H., & Noctor, G. (2011). Ascorbate and glutathione: the heart of the redox hub. Plant Physiology, 155(1), 2-18.

Foyer, C. H., & Noctor, G. (2016). Stress-triggered redox signaling: what's in pROSpect?. Plant, Cell & Environment, 39(5), 951-964.

Foyer, C.H. and Mullineaux, P.M., 1998. The presence of dehydroascorbate and dehydroascorbate reductase in plant tissues. FEBS letters, 425(3), pp.528-529.

Gill, S.S. and Tuteja, N., 2010. Reactive oxygen species and antioxidant machinery in abiotic stress tolerance in crop plants. Plant physiology and biochemistry, 48(12), pp.909-930.

Gómez, L.D., Baud, S., Gilday, A., Li, Y. and Graham, I.A., 2006. Delayed embryo development in the ARABIDOPSIS TREHALOSE-6-PHOSPHATE SYNTHASE 1 mutant is associated with altered cell wall structure, decreased cell division and starch accumulation. The Plant Journal, 46(1), pp.69-84.

González-Schain, N., Dreni, L., Lawas, L.M., Galbiati, M., Colombo, L., Heuer, S., Jagadish, K.S. and Kater, M.M., 2016. Genome-wide transcriptome analysis during anthesis reveals new insights into the molecular basis of heat stress responses in tolerant and sensitive rice varieties. Plant and Cell Physiology, 57(1), pp.57-68.

Haider, S., Raza, A., Iqbal, J., Shaukat, M. and Mahmood, T., 2022. Analyzing the regulatory role of heat shock transcription factors in plant heat stress tolerance: A brief appraisal. Molecular Biology Reports, 49(6), pp.5771-5785.

Hasanuzzaman, M., Nahar, K., Alam, M.M., Roychowdhury, R. and Fujita, M., 2013. Physiological, biochemical, and molecular mechanisms of heat stress tolerance in plants. International journal of molecular sciences, 14(5), pp.9643-9684.

Hatfield, J.L., Boote, K.J., Kimball, B.A., Ziska, L.H., Izaurralde, R.C., Ort, D., Thomson, A.M. and Wolfe, D., 2011. Climate impacts on agriculture: implications for crop production. Agronomy journal, 103(2), pp.351-370.

Hetherington, A.M. and Woodward, F.I., 2003. The role of stomata in sensing and driving environmental change. Nature, 424(6951), pp.901-908.

Higashi, H., Kato, Y., Fujita, T., Iwasaki, S., Nakamura, M., Nishimura, Y., Takenaka, M. and Shikanai, T., 2021. The pentatricopeptide repeat protein PGR3 is required for the translation of petL and ndhG by binding their 5′ UTRs. Plant and Cell Physiology, 62(7), pp.1146-1155.

Hu, S., Ding, Y. and Zhu, C., 2020. Sensitivity and responses of chloroplasts to heat stress in plants. Frontiers in Plant Science, 11, p.375.

IPCC. (2014). Climate Change 2014: Impacts, Adaptation, and Vulnerability. Part A: Global and Sectoral Aspects. Contribution of Working Group II to the Fifth Assessment Report of the Intergovernmental Panel on Climate Change. Cambridge University Press.

Kollist, H., Zandalinas, S.I., Sengupta, S., Nuhkat, M., Kangasjärvi, J. and Mittler, R., 2019. Rapid responses to abiotic stress: priming the landscape for the signal transduction network. Trends in plant science, 24(1), pp.25-37.

Larkindale, J. and Huang, B., 2005. Effects of abscisic acid, salicylic acid, ethylene and hydrogen peroxide in thermotolerance and recovery for creeping bentgrass. Plant Growth Regulation, 47, pp.17-28.

Lawson, T. and Blatt, M.R., 2014. Stomatal size, speed, and responsiveness impact on photosynthesis and water use efficiency. Plant physiology, 164(4), pp.1556-1570.

Lobell, D. B., Schlenker, W., & Costa-Roberts, J. (2011). Climate trends and global crop production since 1980. Science, 333(6042), 616-620.

Lobell, D.B. and Gourdji, S.M., 2012. The influence of climate change on global crop productivity. Plant physiology, 160(4), pp.1686-1697.

Mène-Saffrané, L. and DellaPenna, D., 2010. Biosynthesis, regulation and functions of tocochromanols in plants. Plant Physiology and Biochemistry, 48(5), pp.301-309.

Merilo, E., Jalakas, P., Laanemets, K., Mohammadi, O., Hõrak, H., Kollist, H. and Brosché, M., 2015. Abscisic acid transport and homeostasis in the context of stomatal regulation. Molecular plant, 8(9), pp.1321-1333.

Mhamdi, A., Queval, G., Chaouch, S., Vanderauwera, S., Van Breusegem, F. and Noctor, G., 2010. Catalase function in plants: a focus on Arabidopsis mutants as stress-mimic models. Journal of experimental botany, 61(15), pp.4197-4220.

Mignolet-Spruyt, L., Xu, E., Idänheimo, N., Hoeberichts, F.A., Mühlenbock, P., Brosché, M., Van Breusegem, F. and Kangasjärvi, J., 2016. Spreading the news: subcellular and organellar reactive oxygen species production and signaling. Journal of experimental botany, 67(13), pp.3831-3844.

Mittler, R. (2017). ROS are good. Trends in plant science, 22(1), 11-19.

Mittler, R., 2002. Oxidative stress, antioxidants and stress tolerance. Trends in plant science, 7(9), pp.405-410.

Mittler, R., 2006. Abiotic stress, the field environment and stress combination. Trends in plant science, 11(1), pp.15-19.

Murata, N., Takahashi, S., Nishiyama, Y. and Allakhverdiev, S.I., 2007. Photoinhibition of photosystem II under environmental stress. Biochimica et Biophysica Acta (BBA)-Bioenergetics, 1767(6), pp.414-421.

Nag, A., King, S. and Jack, T., 2009. miR319a targeting of TCP4 is critical for petal growth and development in Arabidopsis. Proceedings of the National Academy of Sciences, 106(52), pp.22534-22539.

Noctor, G., Mhamdi, A. and Foyer, C.H., 2016. Oxidative stress and antioxidative systems: recipes for successful data collection and interpretation. Plant, cell & environment, 39(5), pp.1140-1160.

Noctor, G., Mhamdi, A., Chaouch, S., Han, Y.I., Neukermans, J., Marquez-Garcia, B.E.L.E.N., Queval, G. and Foyer, C.H., 2012.

Glutathione in plants: an integrated overview. Plant, cell & environment, 35(2), pp.454-484.

Noctor, G., Veljovic-Jovanovic, S.O.N.J.A., Driscoll, S., Novitskaya, L. and Foyer, C.H., 2002. Drought and oxidative load in the leaves of C3 plants: a predominant role for photorespiration?. Annals of Botany, 89(7), pp.841-850.

Nouri, M.Z., Moumeni, A. and Komatsu, S., 2015. Abiotic stresses: insight into gene regulation and protein expression in photosynthetic pathways of plants. International journal of molecular sciences, 16(9), pp.20392-20416.

Pandey, K.B. and Rizvi, S.I., 2009. Plant polyphenols as dietary antioxidants in human health and disease. Oxidative medicine and cellular longevity, 2, pp.270-278.

Pandey, P., Irulappan, V., Bagavathiannan, M.V. and Senthil-Kumar, M., 2017. Impact of combined abiotic and biotic stresses on plant growth and avenues for crop improvement by exploiting physio-morphological traits. Frontiers in plant science, 8, p.537.

Per, T.S., Khan, M.I.R., Anjum, N.A., Masood, A., Hussain, S.J. and Khan, N.A., 2018. Jasmonates in plants under abiotic stresses: Crosstalk with other phytohormones matters. Environmental and experimental botany, 145, pp.104-120.

Pignocchi, C., Kiddle, G., Hernández, I., Foster, S.J., Asensi, A., Taybi, T., Barnes, J. and Foyer, C.H., 2006. Ascorbate oxidase-dependent changes in the redox state of the apoplast modulate gene transcript accumulation leading to modified hormone signaling and orchestration of defense processes in tobacco. Plant physiology, 141(2), pp.423-435.

Pospíšil, P., 2009. Production of reactive oxygen species by photosystem II. Biochimica et Biophysica Acta (BBA)-Bioenergetics, 1787(10), pp.1151-1160.

Prasad, P.V. and Djanaguiraman, M., 2011. High night temperature decreases leaf photosynthesis and pollen function in grain sorghum. Functional Plant Biology, 38(12), pp.993-1003.

Prasad, P.V.V., Boote, K.J., Allen Jr, L.H., Sheehy, J.E. and Thomas, J.M.G., 2006. Species, ecotype and cultivar differences in spikelet fertility and harvest index of rice in response to high temperature stress. Field crops research, 95(2-3), pp.398-411.

Rai, R., Singh, S., Rai, S., Shankar, A., Chatterjee, A. and Rai, L.C., 2019. Abiotic Stress-Mediated Oxidative Damage in Plants: An Overview. Approaches for enhancing abiotic stress tolerance in plants, pp.227-252.

Raschke, K., 1975. Stomatal action. Annual Review of Plant Physiology, 26(1), pp.309-340.

Ruan, Y.L., Jin, Y., Yang, Y.J., Li, G.J. and Boyer, J.S., 2010. Sugar input, metabolism, and signaling mediated by invertase: roles in development, yield potential, and response to drought and heat. Molecular Plant, 3(6), pp.942-955.

Sharkey, T.D., 1988. Estimating the rate of photorespiration in leaves. Physiologia Plantarum, 73(1), pp.147-152.

Sharma, P., Jha, A.B., Dubey, R.S. and Pessarakli, M., 2012. Reactive oxygen species, oxidative damage, and antioxidative defense mechanism in plants under stressful conditions. Journal of botany, 2012.

Shikanai, T., 2007. Cyclic electron transport around photosystem I: genetic approaches. Annu. Rev. Plant Biol., 58, pp.199-217.

Sonoike, K., 2011. Photoinhibition of photosystem I. Physiologia plantarum, 142(1), pp.56-64.

Suzuki, N., Koussevitzky, S.H.A.I., Mittler, R.O.N. and Miller, G.A.D., 2012. ROS and redox signaling in the response of plants to abiotic stress. Plant, cell & environment, 35(2), pp.259-270.

Takahashi, S. and Murata, N., 2008. How do environmental stresses accelerate photoinhibition?. Trends in plant science, 13(4), pp.178-182.

Tracz-Gaszewska, Z., Klimczak, M., Biecek, P., Herok, M., Kosinski, M., Olszewski, M.B., Czerwińska, P., Wiech, M., Wiznerowicz, M., Zylicz, A. and Zylicz, M., 2017. Molecular chaperones in the acquisition of cancer cell chemoresistance with mutated TP53 and MDM2 up-regulation. Oncotarget, 8(47), p.82123.

Triantaphylidès, C. and Havaux, M., 2009. Singlet oxygen in plants: production, detoxification and signaling. Trends in plant science, 14(4), pp.219-228.

Verslues, P.E. and Zhu, J.K., 2007. New developments in abscisic acid perception and metabolism. Current Opinion in Plant Biology, 10(5), pp.447-452.

Vierling, E., 1991. The roles of heat shock proteins in plants. Annual review of plant biology, 42(1), pp.579-620.

Wahid, A., Gelani, S., Ashraf, M., & Foolad, M. R. (2007). Heat tolerance in plants: An overview. Environmental and Experimental Botany, 61(3), 199-223.

Wang, W., Vinocur, B., Shoseyov, O. and Altman, A., 2004. Role of plant heat-shock proteins and molecular chaperones in the abiotic stress response. Trends in plant science, 9(5), pp.244-252.

Wei, L.J., Deng, X.G., Zhu, T., Zheng, T., Li, P.X., Wu, J.Q., Zhang, D.W. and Lin, H.H., 2015. Ethylene is involved in brassinosteroids induced alternative respiratory pathway in cucumber (Cucumis sativus L.) seedlings response to abiotic stress. Frontiers in Plant Science, 6, p.982.

Yamamoto, Y., 2016. Quality control of photosystem II: the mechanisms for avoidance and tolerance of light and heat stresses are closely linked to membrane fluidity of the thylakoids. Frontiers in Plant Science, 7, p.1136.

Yamori, W., Hikosaka, K. and Way, D.A., 2014. Temperature response of photosynthesis in C 3, C 4, and CAM plants: temperature acclimation and temperature adaptation. Photosynthesis research, 119, pp.101-117.

Yang, B., Tang, J., Yu, Z., Khare, T., Srivastav, A., Datir, S. and Kumar, V., 2019. Light stress responses and prospects for engineering light stress tolerance in crop plants. Journal of Plant Growth Regulation, 38, pp.1489-1506.

# CHAPTER 3

# SENSING OF HEAT STRESS BY PLANTS


## Abstract

Heat stress sensing/signaling in plants involves a complex network of cellular and molecular events that enable them to sense and respond to HT. Heat stress in plants can harm their growth, development and productivity. Fortunately, plants have various mechanisms to sense and respond to high temperatures. Plants possess specialised receptors that can detect temperature changes, often proteins or ion channels located on the plant cell membranes. When temperatures exceed the optimum threshold, these receptors trigger signaling pathways that initiate heat stress responses. Calcium ions play a crucial role in signaling within plant cells. Elevated temperatures can increase intracellular calcium levels, which acts as a signal for heat stress. ROS are signaling molecules that can activate specific genes in heat stress responses. This chapter describes the heat stress sensing mechanism plants deploy to cope with heat stress.



**Keywords**: Sensing, Small RNAs, miRNA, Protein kinase, CDPK


## Introduction

Plants have evolved various mechanisms to sense and respond to heat stress. While they do not possess specialized organs like animals, they have developed intricate molecular and physiological processes to detect and react to high temperatures (Rikhvanov et al., 2014; Li et al., 2014). In addition, plants have temperature receptors distributed throughout their cells. These receptors can detect changes in ambient temperature and initiate responses accordingly. They are sensitive to both gradual and rapid changes in temperature. For example, Transient Receptor Potential (TRP) channels are ion channels that are found in plant cells and are involved in temperature sensing (Rikhvanov et al., 2014). They are sensitive to temperature changes and can be activated or deactivated by specific temperature thresholds. TRP channels have been implicated in mediating thermotropism (directional growth in response to temperature), heat stress responses, and other temperature-dependent processes in plants (Rikhvanov et al., 2014).

Specific ion channels in the plasma membrane of plant cells can respond to temperature changes. For example, the cyclic nucleotide-gated channels (CNGCs) are involved in temperature perception in *Arabidopsis thaliana* (Donnelly, 2017). These channels allow the influx or efflux of ions, leading to changes in membrane potential and subsequent signal transduction pathways (Donnelly, 2017). Phytochromes are photoreceptor proteins that play a role in various light-dependent processes in plants (Franklin and Whitelam, 2005). However, they can also perceive temperature changes. Phytochrome B (PHYB) is known to be involved in temperature sensing and regulates thermo-morphogenesis, which is the growth and development of plants in response to temperature cues (Legris et al., 2016).

Calcium ions play a crucial role in signaling pathways within plants. Heat stress increases cytosolic calcium levels, which acts as a secondary messenger. This calcium signal is sensed by specific proteins, such as calcium sensors, which then transmit the signal to various downstream targets (Dodd et al., 2010). Plants possess particular proteins known as calcium sensors or calcium-binding proteins that can perceive changes in cytosolic calcium levels (Wang et al., 2014). These sensors include calcium-dependent protein kinases (CDPKs), calmodulins (CaMs), and calmodulin-like proteins (CMLs).

The elevated calcium concentration during heat stress binds to these sensors, leading to their activation and initiation of signaling cascades (Wang et al., 2014). Calcium signaling pathways activated by heat stress are involved in regulating the expression of heat stress-responsive genes (Dodd et al., 2010). Calcium-dependent transcription factors, such as members of the $C_2H_2$ zinc finger transcription factor family, are activated by calcium signals and bind to specific DNA sequences to modulate gene expression (Dodd et al., 2010). These transcription factors regulate the synthesis of heat shock proteins (HSPs), antioxidants, and other protective molecules that help plants cope with heat stress (Dodd et al., 2010; Fragkostefanakis et al., 2015).

High temperatures can cause the denaturation of proteins, leading to their unfolding and loss of function (Timperio et al., 2008). Plants have mechanisms to detect unfolded proteins and initiate stress responses. HSPs are specialized proteins that help refold or degrade denatured proteins, preventing cellular damage (Timperio et al., 2008). It's important to note that the understanding of temperature receptors in plants is still an active area of research, and the full range of receptors and their mechanisms are not yet fully elucidated (Hayes et al., 2021). The complexity of plant responses to temperature suggests the involvement of multiple receptor types and intricate signaling networks. Ongoing studies uncover new insights into how plants perceive and respond to temperature cues.

## Types of heat stress response in plants

Plants have evolved various heat stress response mechanisms to mitigate the adverse effects of high temperatures (Table 1). These responses can be categorized into several types, each serving different purposes. Here are some of the common types of heat stress responses in plants. Heat stress influences plant growth and development. Some plants exhibit thermomorphogenesis, which is a specific growth response to high temperatures (Wigge, 2013). It involves alterations in stem elongation, leaf morphology, flowering time, and other developmental processes (Reis et al., 2023). These responses help plants optimize their growth and reproduction under heat-stress conditions. Heat stress can impact plant metabolism, altering various biochemical pathways (Zandalinas et al., 2020). Plants may adjust their metabolic processes to cope with heat stress, for example through modulating photosynthesis, respiration, and carbohydrate metabolism (Hasanuzzaman et al., 2021). These adjustments help redirect resources and energy towards stress mitigation and survival.

Correspondingly, heat stress affects stomatal conductance, leading to increased water loss through transpiration. Plants regulate the stomatal aperture by reducing stomatal opening and increasing stomatal closure to conserve water. This helps to minimize water loss and maintain cellular hydration (Kumar and Wani, 2020). Heat stress can disrupt relative water potential in plants, leading to cellular dehydration and membrane damage. In response, plants undergo osmotic adjustment by accumulating compatible solutes or osmoprotectants. These compounds, such as proline, betaine, and sugars, help maintain cellular water potential, stabilize proteins, and protect cells from dehydration (Kumar and Wani, 2020). High temperatures can lead to the production of reactive oxygen species (ROS) within plant cells. ROS can cause oxidative damage to various cellular components (Siddiqui et al., 2022). To counteract this, plants activate antioxidant defense systems, including enzymes like superoxide dismutase (SOD), catalase, and peroxidases, as well as non-enzymatic antioxidants such as ascorbate (vitamin C), glutathione, and carotenoids (Kumar and Wani, 2020). These antioxidants help scavenge ROS and protect cells from stress-induced oxidative damages.

Certain plant hormones play pivotal roles in heat stress responses. Abscisic acid (ABA) is the most obvious candidate for sensing heat stress as its levels typically increase under heat stress conditions (Cheikh and Jones, 1994). Under stress conditions, ABA regulates stomatal movement, reduces transpiration, and enhances plant tolerance to heat stress by modulating electrolytic homeostasis (Kumar and Wani, 2020). Other hormones, such as ethylene, salicylic acid, and jasmonic acid, are also involved in heat stress responses, regulating gene expression and signaling pathways (Xiang et al., 2013). Heat stress triggers the synthesis and accumulation of a group of proteins called HSPs (Mayer and Bukau, 2005). HSPs act as molecular chaperones, assisting in proper folding and stabilizing proteins under heat stress conditions. They prevent protein denaturation and aggregation, thus maintaining cellular functions (Hasanuzzaman et al., 2013).

**Table 1** Different mechanisms and signaling pathways involved in sensing heat stress tolerance in plants

| Mechanism/Pathway | Functional mechanism | Signaling Molecules/Proteins | References |
|---|---|---|---|
| Calcium Signaling | Calcium ions serve as crucial second messengers in heat stress responses, mediating various downstream signaling events. | Calmodulin (CaM), Calcineurin B-like proteins (CBLs), Calcium-dependent protein kinases (CDPKs) | Knight (1999) |
| Mitogen-Activated Protein Kinase (MAPK) Signaling | MAPK cascades transmit heat stress signals, leading to the activation of downstream target genes involved in stress tolerance. | MAP Kinase Kinases (MKKs), MAP Kinases (MAPKs), MAPK Kinase Kinases (MAPKKKs) | Cristina et al. (2010) |
| Heat Shock Transcription Factors (HSFs) | HSFs are key regulators of the heat stress response, | Heat Shock Factor 1 (HSF1), Heat Shock Factor 2 (HSF2) | Kotak et al. (2007) |

| | | | |
|---|---|---|---|
| | activating the transcription of heat shock proteins and other protective genes. | | |
| Reactive Oxygen Species (ROS) Signaling | ROS function as signaling molecules in heat stress responses, triggering various defense mechanisms and adaptive responses. | Hydrogen Peroxide ($H_2O_2$), Superoxide Anion ($O_2^{\bullet -}$), Singlet Oxygen ($^1O_2$), Hydroxyl Radical (•OH) | Suzuki et al. (2012) |
| Phytohormone Signaling | Phytohormones regulate heat stress responses and coordinate plant growth and development under stressful conditions. | Abscisic Acid (ABA), Salicylic Acid (SA), Jasmonic Acid (JA), Ethylene (ET), Brassinosteroids (BRs), Cytokinins (CKs) | Larkindale and Huang (2004) |
| Ubiquitin-Proteasome System (UPS) | The UPS is involved in the degradation of misfolded proteins and regulates the stability of heat stress-related proteins. | Ubiquitin Ligases (E3), Ubiquitin (Ub), 26S Proteasome | Pratt (1997) |
| Heat Shock Proteins (HSPs) | HSPs are molecular chaperones that assist in protein folding and protect cellular structures under heat stress conditions. | HSP70, HSP90, HSP100, Small HSPs (sHSPs) | Li et al. (2020) |
| RNA-Binding Proteins | RNA-binding proteins are involved in post-transcriptional regulation, RNA processing, and stability under heat stress. | Heat Stress-Inducible Factors (HSFs), Cold-Inducible RNA-Binding Proteins (CIRBPs), RNA-Binding Proteins (RBPs) | Narsai et al. (2011) |

## The complexity of heat stress response in plants

The heat stress response in plants is a complex phenomenon involving various physiological, biochemical, and molecular mechanisms (Table 2) (Mittler et al., 2012). The complexity of heat stress response depends mainly on the intensity and duration of the heat stress, as well as the plant species and its developmental stage (Suzuki et al., 2014). The first and foremost heat stress response is perceived at the cellular and molecular level, triggering a cascade of cellular and molecular events in plants (Mittler et al., 2012). It affects the integrity of cellular membranes, leading to the leakage of ions and other molecules. ROS production is increased, causing oxidative stress (Mittler et al., 2004). HSPs are synthesized to protect proteins from denaturation and also act as molecular chaperones, thus activating other signaling molecules and transcription factors to regulate the expression of defense-responsive genes/proteins (Hemantaranja et al., 2014).

Studies have shown that HSFs also play a crucial role in heat sensing and the subsequent heat stress response in plants (Nover et al., 1996). HSFs are a family of transcription factors regulating heat stress-responsive genes' expression. Under normal conditions, HSFs are typically present in an inactive state in the cytoplasm, forming multimeric complexes with chaperones (Xiang et al., 2013). When plants are exposed to heat stress, the elevated temperature disrupts these complexes, activating HSFs. Once activated, HSFs undergo conformational changes, allowing them to form trimers, which is a crucial step for HSF activation and its corresponding DNA binding functions (Wang et al., 2016). The trimeric HSFs are then translocated into the nucleus, where they exert their regulatory functions after binding to specific DNA sequences called heat shock elements (HSEs) present in the promoter regions of heat stress-responsive genes (Bharti et al., 2004). HSEs are recognized by the consensus sequence 5'-nGAAnnTTCn-3', where 'n' represents any nucleotide. HSF binding to HSEs activates the transcription of HSPs, which in turn modulate the expression of other target genes (Nover et al., 1996). Furthermore, studies have also confirmed that HSFs act as masters of transcriptional regulation of a wide range of heat stress-responsive genes, including molecular chaperones, HSPs, and other protective proteins (Wang et al., 2016). These genes encode proteins that assist in protein folding, prevent protein aggregation, and facilitate the removal of damaged proteins (Xiang et al., 2013).

Several post-translational modifications, such as acetylation, phosphorylation, ubiquitination, and biotinylation, decisively affect HSF activity and stability (Charng et al., 2007). Depending on the specific residues targeted, these modifications either enhance or repress their transcriptional activity. Acetylation, on the other hand, generally enhances HSF activity (Mishra et al., 2002). HSFs are involved in the development of heat stress memory in plants, allowing them to respond more effectively to recurring heat stress episodes (Lämke and Bäurle, 2017). Heat stress-induced activation of HSFs leads to the transcriptional upregulation of genes involved in heat stress memory, which helps plants develop a more robust and faster response upon subsequent encounter with heat stress conditions (Ding et al., 2012).

**Table 2** Complexity and types of heat stress responses in plants, including the role of unfolded protein response (UPR) in sensing heat stress

| Types of Heat Stress Response | Complex response | Role of Unfolded Protein Response (UPR) in Sensing Heat Stress | References |
|---|---|---|---|
| Molecular Complexity of Heat Stress Response | The heat stress response in plants involves intricate molecular mechanisms and interactions that regulate gene expression, protein stability, and metabolic pathways to maintain cellular homeostasis and promote stress tolerance. | The UPR is a conserved signaling pathway that is activated under heat stress conditions. It activates chaperones, protein-folding enzymes, and degradation systems to assist in protein folding and quality control. | Fanata et al. (2013) |
| Transcriptional Responses | Heat stress triggers extensive transcriptional reprogramming of various heat stress-responsive genes. This transcriptional response includes the upregulation of heat shock | The UPR in plants involves the activation of specific transcription factors, such as bZIP17 and bZIP28, which regulate the expression of UPR-related genes. These TFs initiate downstream signaling pathways that modulate the expression of target genes involved in stress response. | Iwata et al. (2017); Ruberti et al. (2015) |

| | | | |
|---|---|---|---|
| | proteins, chaperones, proteases, antioxidant enzymes, and other stress-related genes. | | |
| Protein Quality Control Mechanisms | Heat stress disrupts protein folding, leading to the accumulation of unfolded or misfolded proteins. Plants employ various mechanisms including molecular chaperones, heat shock proteins, proteases, and autophagy to degrade them | The UPR coordinates protein quality control mechanisms in the ER under heat stress. It activates chaperones, such as binding immunoglobulin protein (BiP), to assist in protein folding and prevent protein aggregation. The UPR also upregulates ER-associated degradation (ERAD) components, including E3 ubiquitin ligases and ERAD-specific chaperones, to selectively target misfolded proteins for degradation. | Manghwar and Li, (2022) |
| Metabolic Adjustments | Heat stress alters metabolic pathways in plants, including changes in energy metabolism, carbon partitioning, and the synthesis of compatible solutes and protective compounds. | The UPR influences metabolic adjustments under heat stress. It regulates the synthesis of protective compounds, such as heat shock proteins and antioxidant enzymes, to enhance stress tolerance. | Manghwar and Li, (2022) |

## Sensing of HT stress by plants

In various organisms, the plasma membrane acts as an inception site for sensing heat stress (Saidi et al., 2011). Several scientists have used biochemical measurements in conjunction with electrophysiology and reporter gene assays, thus corroborating the above notion (Hasanuzzaman et al., 2013; Szechyńska-Hebda et al., 2022). Their experiments revealed that slight temperature variations are sensed primarily at the plasma membrane, and the signal is then perceived by calcium channels, thus triggering calcium influx and activating other signaling molecules involved in the perception of heat stress response (Figure 1) (Scharf et al., 2012; Bita and Gerats, 2013). The plasma membrane of plant cells plays a crucial role in responding to and coping with heat stress. Heat stress affects the integrity and functionality of the plasma membrane, which is the primary site of temperature perception in plants. Here are some critical aspects of the plasma membrane and heat stress interaction.

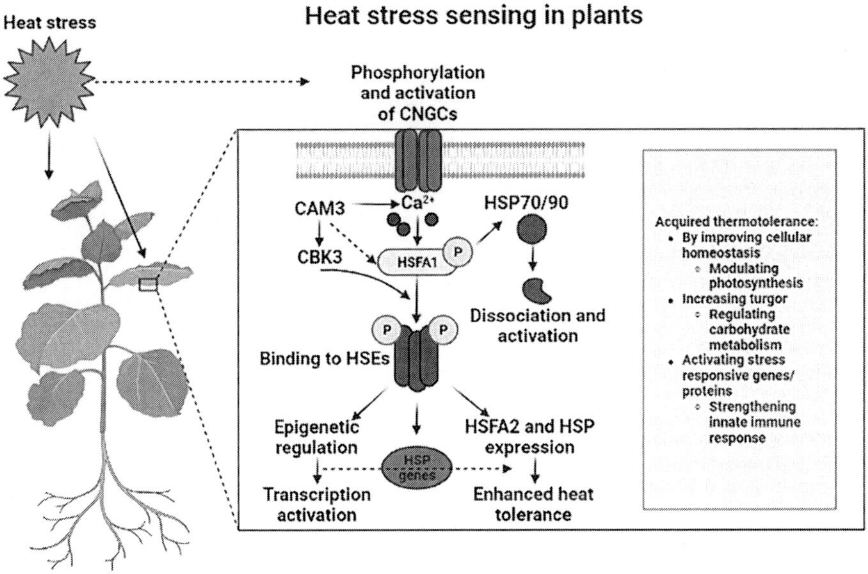


**Figure 1** Mechanism of heat stress signal perception and induction of thermotolerance in plants. High temperature alter the fluidity of plasma memebrane leading to phosphorylation and activation of calcium channels. Transient entry of $Ca^{2+}$ triggers heat shock factors (HSF) signaling cascades activating HSP and other stress responsive genes leading to increase HT stress tolerance.

## Plasma membrane-mediated heat sensing

Heat stress is known to disrupt the lipid bilayer of the plasma membrane, altering its fluidity (Liu et al., 2006). At high temperatures, lipids become more fluid, potentially causing membrane leakage and compromising the membrane's selective permeability. Changes in membrane fluidity can impact the activity and stability of membrane proteins and other components embedded within the plasma membrane (Wang et al., 2017). Heat stress disrupts ion homeostasis by affecting ion channels and transporters embedded in the plasma membrane. Elevated temperatures can alter the gating potential of ion channels, tremendously increasing ion fluxes across the membrane (Richter et al., 2010). This disturbance in ion homeostasis can disrupt cellular processes and lead to cellular damage. The plasma membrane is a central hub for discrete stress defense response signaling events. Changes in membrane potential and ion homeostasis regulate calcium signaling pathways, leading to the activation of downstream heat stress responses (Wang et al., 2004). Heat stress also triggers the production of lipid-derived signaling molecules, such as phosphatidic acid, which regulate stress signaling cascades (Kim et al., 2010).

The plasma membrane also has the potential for its repair and remodeling under heat stress, and plants do it by exaggerating the synthesis of lipids and simultaneously removing the damaged lipids, thus strengthening the integrity of the membrane (Vera-Estrella et al., 1994). Additionally, lipid remodeling can occur to adjust the lipid composition and maintain membrane stability under heat stress conditions (Khan et al., 2023). Calcium ($Ca^{2+}$) channels play a crucial role in the response of plants to heat stress. They are involved in perceiving and transmitting heat stress signals, as well as triggering downstream cellular and molecular responses (Chinnusamy et al., 2009). Calcium channels are implicated in the perception of high temperatures by plants. Heat stress activates temperature-sensitive calcium channels, which influx calcium ions ($Ca^{2+}$) into the cytosol. This increase in cytosolic calcium serves as an early signal in the heat stress response pathway (Batistič and Kudla, 2012). The influx of calcium ions triggers a complex signaling cascade, known as calcium signaling, which involves calcium sensors and downstream signaling components (Tezuka et al., 2021). Calcium-binding proteins, such as calmodulins (CaMs) and calcineurin B-like proteins (CBLs), sense the changes in cytosolic calcium levels and transduce the signal to downstream targets (Wang et al., 2014).

Calcium signaling activates various downstream defense-responsive genes/proteins via activating Calcium-dependent protein kinases (CDPKs) (Tezuka et al., 2021). They phosphorylate transcription factors, such as HSFs, and other regulatory proteins, leading to the activation of heat stress-responsive genes. This triggers the synthesis of HSPs, antioxidant enzymes, and other protective molecules (Saidi et al., 2009). Calcium channels are involved in maintaining ion homeostasis during heat stress. Elevated temperatures can disrupt ion gradients and induce ion fluxes (Hepler et al., 2012). Calcium channels, together with other ion channels and transporters, play a role in balancing intracellular ion concentrations, particularly calcium ions, to mitigate the adverse effects of heat stress (Pirayesh et al., 2021). Understanding the role of calcium channels in heat stress responses is essential for deciphering the molecular mechanisms underlying plant thermotolerance. By mediating calcium signaling and activating downstream responses, calcium channels contribute to coordinating heat stress defense mechanisms in plants, including gene expression regulation, stomatal responses, and ion homeostasis maintenance.

## Unfolded protein response for heat sensing

The unfolded protein response (UPR) is a cellular mechanism that is activated in the endoplasmic reticulum (ER) under stress conditions (Schröder and Kaufman, 2005). The ER is an organelle involved in protein synthesis, folding, and transport. Various cellular/physiological processes such as nutrient deprivation, viral infection, accumulation of misfolded proteins, or chemical insults can disrupt ER homeostasis (Bao et al., 2019). The above turbulence in the ER can lead to an accumulation of unfolded or misfolded proteins within the ER lumen. In such situations, the UPR is triggered to restore ER function and maintain cellular viability (Schröder and Kaufman, 2005). UPR is a novel stress defense mechanism identified only recently in plants and has been recognized as an important intracellular event (Bao et al., 2019). The UPR functions by stimulating ER proteostasis, which stimulates gene expression in the nucleus to reduce ER stress (Ahammed et al., 2016). Although the functional mechanism of UPR in ameliorating ER stress is more well understood in animals as compared to plants (Bao et al., 2019). Overall, the UPR attempts to process the accumulation of unfolded proteins by downregulating the transcription of secretory proteins and increases the removal of misfolded or slowly folding proteins through ER-associated degradation (ERAD) or lysosomal degradation (Ruberti et al., 2015).

In animals, the UPR is mediated by three main transmembrane proteins located in the ER membrane: inositol-requiring enzyme 1 (IRE1), protein kinase R-like ER kinase (PERK), and activating transcription factor 6

(ATF6). These sensors detect the accumulation of unfolded proteins and initiate a series of signaling events to restore ER homeostasis (Hetz et al., 2019). In plants, one of the primary transmembrane sensors involved in plant UPR is the protein kinase/endoribonuclease IRE1 (Iwata et al., 2017). Upon activation, IRE1 undergoes autophosphorylation and activates its ribonuclease activity. This activity leads to the unconventional splicing of a specific mRNA encoding a transcription factor called bZIP60 (Moreno et al., 2012). The spliced form of bZIP60 is translated into an active transcription factor that regulates the expression of UPR target genes involved in protein folding, ER-associated degradation, and lipid metabolism (Noh and Kwon, 2002).

Another vital sensor in plant UPR is the transmembrane protein bZIP17/bZIP28. These proteins are retained in the ER under normal conditions but are released upon ER stress. Once released, bZIP17/bZIP28 translocate to the nucleus and activate the expression of UPR-related genes (Liu and Howell, 2010). Additionally, other transcription factors such as NAC (NAM, ATAF1/2, CUC2) and basic leucine zipper (bZIP) family members, including bZIP60 and bZIP28, play crucial roles in regulating UPR gene expression and coordinating the plant's response to ER stress (Nakashim et al., 2012; Servet et al., 2010). The UPR in plants also involves other components, such as protein disulfide isomerases (PDIs), chaperones, and components of the ER-associated degradation (ERAD) pathway. These proteins assist in protein folding, quality control, and degradation of misfolded proteins to restore ER homeostasis (Schröder and Kaufman 2005; Hüttner and Strasser, 2012). The UPR in plants is an essential cellular response to ER stress caused by various environmental and developmental factors. It involves the activation of specific signaling pathways and transcription factors to restore ER homeostasis and ensure plant survival under adverse conditions.

## Metabolic sensors and ROS signaling

Heat stress is a significant environmental factor that can adversely affect plant growth, development, and productivity (Wahid et al., 2007). To cope with heat stress, plants have evolved various mechanisms, including the activation of metabolic processes (Bita and Gerats, 2013). While there are no specific metabolic sensors for heat stress in plants, plants possess specific metabolic pathways and associated enzymes that play crucial roles in the heat stress response (Nover et al., 1996). These pathways and enzymes can indirectly serve as indicators or markers of plant heat stress.

One such metabolic marker/sensor is the peroxidation of membrane lipids. Heat stress can cause the peroxidation of membrane lipids in plants. The breakdown of lipids generates malondialdehyde (MDA) and other reactive aldehydes, which can be quantified as an indicator of heat stress-induced lipid peroxidation (Sharma et al., 2012). Oxidation of membrane lipids under heat stress is triggered by the production of ROS in plant cells. ROS, including superoxide radicals and hydrogen peroxide, are generated primarily in chloroplasts, mitochondria, and peroxisomes (Suzuki et al., 2012).

Among all, hydroxyl radicals ($^{\bullet}OH$) is one of the most attacking ROS, posing a severe threat to polyunsaturated fatty acids (PUFAs) present in membrane lipids (Apel and Hirt, 2004). The presence of multiple double bonds makes PUFAs highly susceptible to oxidative damage. ROS initiate a chain of oxidation reactions by abstracting a hydrogen atom from a PUFA, thus generating a lipid radical ($L^{\bullet}$) and a lipid peroxyl radical ($LOO^{\bullet}$) (Tripathy and Oelmüller, 2012). The generated lipid ($L^{\bullet}$) and ($LOO^{\bullet}$) then react with another PUFA molecule resulting in the formation of lipid hydroperoxide (LOOH) and other lipid radicals (Tripathy and Oelmüller, 2012). This process propagates the chain reaction, and lipid hydroperoxides accumulate within the membrane, which upon their subsequent decomposition, forms a variety of reactive aldehydes, including MDA (Tiwari et al., 2019). Once generated, these MDA and other reactive aldehydes rapidly accumulate under heat stress, altering the physical properties of membranes, including fluidity, permeability, and stability (Mittler et al., 2004). This disruption can compromise membrane functions and lead to cellular damage (Mittler et al., 2004).

Proline is an amino acid that accumulates in plants under stress conditions, including heat stress (Hayat et al., 2012). Increased proline levels can be considered a metabolic marker of heat stress, serving as osmoprotectants that protect membrane lipids from oxidative damage (Szabados and Savouré, 2010). In addition, proline accumulation maintains osmotic balance and prevents dehydration in plants exposed to heat stress. Increased proline concentration upon stress exposure helps plants retain water and maintain turgor pressure, thus stabilizing membrane integrity (Hayat et al., 2012). Increased accumulation of proline is an excellent indicator of heat stress severity (Kazerooni et al., 2021). However, it is essential to note that proline accumulation can also occur in response to other stresses or developmental processes. Therefore, it is necessary to consider proline levels in conjunction with other physiological and biochemical indicators to comprehensively evaluate plant responses to heat stress.

## Molecular changes occurring in response to HT stress

Heat stress can have profound effects on plant physiology, with concomitant changes at cellular and molecular levels. When plants are exposed to high temperatures, a series of molecular responses are activated, which are aimed to protect them from the damaging effects of heat stress-induced oxidative damages (Bokszczanin and Fragkostefanakis, 2015). Heat stress triggers changes in intracellular calcium concentrations. Calcium ions ($Ca^{2+}$) act as potent secondary messengers, activating various cellular processes (Mittler et al., 2012). Enhanced calcium signaling modulates specific proteins and signaling pathways involved in stress defense responses, including the expression of heat shock proteins and other stress-related genes (Hasanuzzaman et al., 2013). Furthermore, heat stress induces changes in the expression pattern of enzymatic and non-enzymatic antioxidants and proteins involved in osmotic regulation (Mittler et al., 2012). The calcium signaling also activates a negative feedback loop by downregulating the expression of genes associated with growth and development to conserve energy and resources (Mittler et al., 2012).

## Histone modification

Histone modifications play a crucial role in regulating gene expression in response to various environmental stimuli, including heat stress, in plants (Probst et al., 2004). Histone modifications under heat stress trigger plant molecular and physiological responses. In addition, the histone modifications can also contribute to the transcriptional reprogramming, which is essential for systemic rewiring of adaptation and survival under climate extremes (Mastrangelo et al., 2012). Acetylation is the most common histone modification that occurs in response to external stimuli, including heat stress. Several studies have confirmed that histone acetylation is generally associated with the activation of gene expression (Kumar and Wigge, 2010). Acetylation functions by neutralizing the positive charge on histone proteins, causing conformational/structural changes to chromatin structure in such a way that it enhances the accessibility of chromatin transcriptional machinery to gene promoters (Jaskiewicz et al., 2011). The most relevant example of histone acetylation is the heat shock genes, which encode heat shock proteins involved in cellular protection and stress tolerance (Lämke et al., 2016).

Likewise, histone methylation can exert differential effects on the expression of defense-responsive genes/proteins by methylating specific residues of histone proteins (Kim et al., 2010). For example, trimethylation of histone H3 lysine 4 (H3K4me3) is generally associated with active gene expression, while trimethylation of H3 lysine 27 (H3K27me3) is typically linked to gene repression (Valmonte et al., 2014). Heat stress can alter the balance of histone methylation marks, leading to changes in gene expression patterns. Similarly, phosphorylation of histones under stress conditions can affect chromatin structure as well as transcription of various downstream genes (Wang et al., 2014). Phosphorylation of histone H3 serine 10 (H3S10ph) is a conserved mark associated with the activation of gene expression (Jiang et al., 2007). Phosphorylation promotes decondensation of chromatin, thus facilitating efficient binding of transcription factors and other regulatory proteins to specific gene loci (Jiang et al., 2007). Ubiquitination is a post-translational modification that targets unfolded or misfolded proteins for degradation or alters their function (Sharma et al., 2021). Heat stress-induced ubiquitination of histones can lead to changes in gene expression by regulating chromatin accessibility and recruitment of transcriptional regulators (Zhou et al., 2017). These histone modifications collectively contribute to the dynamic regulation of gene expression in plants exposed to HT stress. They coordinatively regulate transcription factors and non-coding RNAs to orchestrate synergistic stress response, thus strengthening plant adaptation and survival under HT stress. However, it is essential to note that the specific histone modifications and their effects can vary depending on the plant species and the severity and duration of the heat stress.

## Chromatin remodelling

Under heat stress, cells activate a plethora of cellular and molecular responses to protect themselves from the detrimental effects of oxidative damage (Mittler et al., 2004). One crucial aspect involves chromatin remodeling to induce cellular adaptation under HT stress. Chromatin remodeling refers to the dynamic conformational changes that modify the packaging and accessibility of DNA to specific genes (Probst et al., 2004; Pecinka et al., 2009). Studies have reported that heat stress often leads to the relaxation or decondensation of chromatin structure. This relaxation/decondensation event instigates better accessibility of DNA to various transcription factors and regulatory proteins (Lämke and Bäurle, 2017). The efficient binding of a gene to its specific promoter activates the expression of defense-responsive genes (Lämke and Bäurle, 2017). Heat stress triggers the addition or removal of specific chemical modifications on histone proteins, which are the integral components of chromatin (Kumar and Wigge, 2010). Acetylation of histones is associated with gene activation, and heat stress can promote histone acetylation at specific genomic regions to enhance the expression of heat shock proteins and various stress signaling genes (Lämke and Bäurle, 2017; Chen et al., 2010). The ATP-dependent chromatin remodeling also plays a crucial role in reorganizing and repositioning nucleosomes, which are the basic units of chromatin (Flaus and Owen-Hughes, 2011).

ATP-mediated hydrolysis alters nucleosome positions, exposes regulatory DNA elements, and modulates gene expression. Under HT stress, these remodeling complexes may be recruited to specific genomic loci to facilitate transcriptional activation or repression of stress-responsive genes. Heat stress often leads to the recruitment of specific transcription factors to HSEs within the promoter regions of stress-responsive genes (Scharf et al., 2012). Chromatin remodeling events can enhance the accessibility of HSEs and facilitate the binding of these transcription factors, promoting gene activation. In addition, chromatin remodeling events, such as relaxation or decondensation, render optimum accessibility of HSEs, thus facilitating efficient binding of HSFs to these regulatory regions (Bharti et al., 2004). Chromatin remodeling events induced by heat stress can leave epigenetic marks that persist even after the pressure is relieved (Lämke and Bäurle, 2017). These epigenetic changes can contribute to cellular memory, enabling a faster and more efficient response to subsequent heat stress episodes.

Chromatin remodeling complexes, such as SWI/SNF and ISWI, also play a significant role in modulating the expression of heat shock protein genes (Song et al., 2021). These ATP-dependent complexes can efficiently enhance the availability of heat shock protein gene loci to the heat shock protein gene promoters, thus contributing to chromatin remodeling events (Luo et al., 2012). In addition, these chromatin modification events are necessary for transcriptional activation, nucleosome repositioning, restructuring, and allowing access to transcriptional machinery to other stress-responsive genes (Kim et al., 2010).

## Post-translational modifications

Several post-translational modifications (PTMs), such as phosphorylation, acetylation, SUMOylation, and others, critically regulate protein function and activate signaling pathways under heat stress in plants (Valmonte et al., 2014; Augustine and Vierstra, 2018). These PTMs have been implicated in modulating plant responses to heat stress, enabling plants' innate immune response for better adaptability and survivability under extreme environments (Qiao and Fan, 2008.). Ubiquitination functions by incorporating ubiquitin molecules to target proteins. This PTM is mainly involved in protein degradation, protein-protein interactions, and subcellular localization (Ling et al., 2019). Further, in plants, ubiquitination-mediated protein degradation explicitly eliminates damaged or misfolded proteins, thus preventing cellular toxicity and maintaining protein homeostasis (Rikhvanov et al., 2014). Another PTM, SUMOylation, regulates protein function by attaching a small ubiquitin-like modifier (SUMO) protein (Castro et al., 2012). It also modulates various cellular and metabolic processes, including stress responses (Xu and Yang, 2013). Under HT stress, SUMOylation of proteins can drastically regulate their activity, stability, and subcellular localization of target proteins, thereby influencing plant stress tolerance (Ling et al., 2019).

Acetylation refers to the addition of an acetyl group to lysine residues in proteins to modulate protein function, stability, and interactions (Zheng, 2020). Under heat stress, levels of protein acetylation can change drastically, affecting the overall functionality of heat-responsive proteins and other downstream signaling genes involved in heat stress tolerance (Zhou et al., 2017). Glycosylation consists of the attachment of sugar molecules to proteins and is an imperative post-translational modification regulating protein folding, stability, and function (Kosová et al., 2018). Heat stress can have detrimental effects on protein glycosylation patterns, influencing their structural and functional integrity. Recently, the role of nitric oxide has come to light in controlling post-translational modifications of target proteins via a process known as nitrosylation (Qiao and Fan 2008.). Nitrosylation functions by adding a nitric oxide (NO) group to cysteine residues of target proteins. NO has long been known to act as a signaling molecule involved in diverse cellular processes, including stress responses (Crawford, 2006). Nitrosylation can also essentially regulate overall protein activity, stability, and signaling cascades to improve the growth and productivity of plants under heat stress (Beligni & Lamattina, 1999). These post-translational modifications are known to significantly regulate protein activities, protein-protein interactions, and signal transduction pathways, allowing plants to mount effective responses to heat stress. By fine-tuning protein functions and cellular processes, post-translational modifications enable plants to cope with high-temperature conditions and enhance their heat stress tolerance.

## Novel components in plant heat stress signaling

In plants, heat stress signaling is a complex process that provokes the activation of various molecular components and signaling pathways (Zandalinas et al., 2020). A large body of literature has documented several essential elements such as calcium channels, phytohormone-mediated signaling, and HSPs that initiate heat stress signaling in plants (Kotak et al., 2007; Ohama et al., 2017). Several biotechnological interventions in plant stress biology have identified and uncovered novel components contributing to this process. This section will highlight some breakthroughs in identifying unknown plant stress signaling components.

Protein kinases are enzymes that phosphorylate proteins and play a vital role in activating stress signaling pathways (Rikhvanov et al., 2014). Recent studies have identified novel protein kinases that efficiently regulate plant heat stress response (Kitsios and Doonan, 2011). For instance, the CDPKs are involved in heat stress signaling in plants. CDPKs function by phosphorylating target proteins, thereby activating the transcription of heat stress-responsive genes (Valmonte et al., 2014).

Small RNAs, including microRNAs (miRNAs) and small interfering RNAs (siRNAs), have been extensively studied for their regulatory roles in post-transcriptional gene expression. Several studies have highlighted the significance of small RNAs in modulating heat stress responses in plants (Khraiwesh et al., 2012; Kantar et al., 2010). Under heat stress conditions, differential expression of miRNAs has been observed, and these miRNAs can target and regulate the expression of heat stress-responsive genes. For example, miR156 is known to regulate the expression of SPL (SQUAMOSA PROMOTER BINDING PROTEIN-LIKE) transcription factors, which play a role in flowering time regulation and heat stress response (Meng et al., 2012; Servet et al., 2010). On the other hand, miR159 has been implicated in regulating MYB (MYELOBLASTOSIS) transcription factors involved in heat stress response and hormone signaling pathways (Sunkar et al., 2012; Li et al., 2016). Additionally, downregulation of miR398 has been shown to regulate the expression of copper superoxide dismutase (CSOD), an antioxidant enzyme involved in scavenging reactive oxygen species

(ROS). Decreased levels of miR398 lead to increased CSOD expression, enhancing the plant's ability to cope with heat-induced oxidative stress (Wang et al., 2016; Zhang et al., 2018).

Similarly, miR319 and miR169 have been implicated in heat stress responses and leaf morphogenesis. The downregulation of miR319 activates TCP (TEOSINTE BRANCHED1, CYCLOIDEA, and PROLIFERATING CELL FACTORS) transcription factors, which promote heat stress tolerance and maintain leaf integrity (Palatnik et al., 2003; Li et al., 2015). Downregulation of miR169 under heat stress activates NFYA (NUCLEAR FACTOR Y, SUBUNIT A) transcription factors, which regulate the expression of heat stress-responsive genes, including heat shock proteins (HSPs) and other stress-related proteins (Vierling, 1991; Han et al., 2020).

## Conclusion

Heat stress signaling in plants evokes an intricate network of cellular and molecular processes that enable them to adapt under high-temperature conditions. MAPK cascades are critical signaling pathways in heat stress responses that phosphorylate and activate MAPK kinases. These activated MAPKs can translocate to the nucleus and phosphorylate transcription factors, leading to the expression of heat stress-responsive genes. Heat stress influences the levels and activity of various phytohormones, which act as signaling molecules. Two essential hormones in heat stress signaling are ABA and ethylene. ABA helps plants regulate stomatal closure, whereas ethylene modulates gene expression and physiological processes. Overall, the signaling mentioned above is a vital mediator in plants' perception of heat stress, transmitting signals, activating specific sensors and transcription factors, and modulating various physiological and biochemical responses to help plants adapt and survive in high-temperature conditions.

## References

Ahammed, G.J., Li, X., Zhou, J., Zhou, Y.H. and Yu, J.Q., 2016. Role of hormones in plant adaptation to heat stress. Plant hormones under challenging environmental factors, pp.1-21.

Apel, K. and Hirt, H., 2004. Reactive oxygen species: metabolism, oxidative stress, and signal transduction. Annu. Rev. Plant Biol., 55, pp.373-399.

Augustine, R.C. and Vierstra, R.D., 2018. SUMOylation: re-wiring the plant nucleus during stress and development. Current opinion in plant biology, 45, pp.143-154.

Bao, Y., Bassham, D.C. and Howell, S.H., 2019. A functional unfolded protein response is required for normal vegetative development. Plant Physiology, 179(4), pp.1834-1843.

Batistič, O. and Kudla, J., 2012. Analysis of calcium signaling pathways in plants. Biochimica et Biophysica Acta (BBA)-General Subjects, 1820(8), pp.1283-1293.

Beligni, M.V. and Lamattina, L., 1999. Is nitric oxide toxic or protective?. Trends in Plant Science, 4(8), pp.299-300.

Bharti, K., von Koskull-Doring, P., Bharti, S., Kumar, P., Tintschl-Korbitzer, A., Treuter, E. and Nover, L., 2004. Tomato heat stress transcription factor HsfB1 represents a novel type of general transcription coactivator with a histone-like motif interacting with the plant CREB binding protein ortholog HAC1. The Plant Cell, 16(6), pp.1521-1535.

Bita, C.E. and Gerats, T., 2013. Plant tolerance to high temperature in a changing environment: scientific fundamentals and production of heat stress-tolerant crops. Frontiers in plant science, 4, p.273.

Bokszczanin, K.L. and Fragkostefanakis, S., 2013. Solanaceae pollen thermotolerance initial training network C: Perspectives on deciphering mechanisms underlying plant heat stress response and thermotolerance. Front. Plant Sci, 4(315), pp.10-3389.

Castro, P.H., Tavares, R.M., Bejarano, E.R. and Azevedo, H., 2012. SUMO, a heavyweight player in plant abiotic stress responses. Cellular and Molecular Life Sciences, 69, pp.3269-3283.

Charng, Y.Y., Liu, H.C., Liu, N.Y., Chi, W.T., Wang, C.N., Chang, S.H. and Wang, T.T., 2007. A heat-inducible transcription factor, HsfA2, is required for extension of acquired thermotolerance in Arabidopsis. Plant physiology, 143(1), pp.251-262.

Cheikh, N. and Jones, R.J., 1994. Disruption of maize kernel growth and development by heat stress (role of cytokinin/abscisic acid balance). Plant physiology, 106(1), pp.45-51.

Chen, L.T., Luo, M., Wang, Y.Y. and Wu, K., 2010. Involvement of Arabidopsi s histone deacetylase HDA6 in ABA and salt stress response. Journal of experimental botany, 61(12), pp.3345-3353.

Chinnusamy, V. and Zhu, J.K., 2009. Epigenetic regulation of stress responses in plants. Current opinion in plant biology, 12(2), pp.133-139.

Crawford, N.M., 2006. Mechanisms for nitric oxide synthesis in plants. Journal of experimental botany, 57(3), pp.471-478.

Cristina, M.S., Petersen, M. and Mundy, J., 2010. Mitogen-activated protein kinase signaling in plants. Annual review of plant biology, 61, pp.621-649.

Deng, Y., Humbert, S., Liu, J.X., Srivastava, R., Rothstein, S.J. and Howell, S.H., 2011. Heat induces the splicing by IRE1 of a mRNA encoding a transcription factor involved in the unfolded protein response in Arabidopsis. Proceedings of the National Academy of Sciences, 108(17), pp.7247-7252.

Ding, Y., Fromm, M. and Avramova, Z., 2012. Multiple exposures to drought'train'transcriptional responses in Arabidopsis. Nature communications, 3(1), p.740.

Dodd, A.N., Kudla, J. and Sanders, D., 2010. The language of calcium signaling. Annual review of plant biology, 61, pp.593-620.

Donnelly, C.C.A., 2017. The role of MOR1 in the control of microtubule organization and dynamics in Arabidopsis thaliana (Doctoral dissertation, University of British Columbia).

Fanata, W.I.D., Lee, S.Y. and Lee, K.O., 2013. The unfolded protein response in plants: a fundamental adaptive cellular response to internal and external stresses. Journal of proteomics, 93, pp.356-368.

Flaus, A. and Owen-Hughes, T., 2011. Mechanisms for ATP-dependent chromatin remodelling: the means to the end. The FEBS journal, 278(19), pp.3579-3595.

Fragkostefanakis, S., Roeth, S., Schleiff, E. and SCHARF, K.D., 2015. Prospects of engineering thermotolerance in crops through modulation of heat stress transcription factor and heat shock protein networks. Plant, cell & environment, 38(9), pp.1881-1895.

Franklin, K.A. and Whitelam, G.C., 2005. Phytochromes and shade-avoidance responses in plants. Annals of botany, 96(2), pp.169-175.

Hasanuzzaman, M., Nahar, K., Alam, M.M., Roychowdhury, R. and Fujita, M., 2013. Physiological, biochemical, and molecular mechanisms of heat stress tolerance in plants. International journal of molecular sciences, 14(5), pp.9643-9684.

Hayat, S., Hayat, Q., Alyemeni, M.N., Wani, A.S., Pichtel, J. and Ahmad, A., 2012. Role of proline under changing environments: a review. Plant signaling & behavior, 7(11), pp.1456-1466.

Hayes, S., Schachtschabel, J., Mishkind, M., Munnik, T. and Arisz, S.A., 2021. Hot topic: Thermosensing in plants. Plant, Cell & Environment, 44(7), pp.2018-2033.Mittler R, Wang Q, Pauwels J, et al. (2022).

Exploring Plant Thermosensing Mechanisms: From Cellular Signaling to Organismal Responses. Plant Cell, 34(1), 37-63. doi: 10.1093/plcell/koab086.

Hemantaranjan, A., Bhanu, A.N., Singh, M.N., Yadav, D.K., Patel, P.K., Singh, R. and Katiyar, D.J.A.P.A.R., 2014. Heat stress responses and thermotolerance. Adv. Plants Agric. Res, 1(3), pp.1-10.

Hepler, P.K., Kunkel, J.G., Rounds, C.M. and Winship, L.J., 2012. Calcium entry into pollen tubes. Trends in plant science, 17(1), pp.32-38.

Hetz, C., Axten, J.M. and Patterson, J.B., 2019. Pharmacological targeting of the unfolded protein response for disease intervention. Nature chemical biology, 15(8), pp.764-775.

Hüttner, S. and Strasser, R., 2012. Endoplasmic reticulum-associated degradation of glycoproteins in plants. Frontiers in plant science, 3, p.67.

Iwata, Y., Ashida, M., Hasegawa, C., Tabara, K., Mishiba, K.I. and Koizumi, N., 2017. Activation of the Arabidopsis membrane-bound transcription factor bZIP 28 is mediated by site-2 protease, but not site-1 protease. The Plant Journal, 91(3), pp.408-415.

Jaskiewicz, M., Conrath, U. and Peterhänsel, C., 2011. Chromatin modification acts as a memory for systemic acquired resistance in the plant stress response. EMBO reports, 12(1), pp.50-55.

Jiang, D., Yang, W., He, Y. and Amasino, R.M., 2007. Arabidopsis relatives of the human lysine-specific Demethylase1 repress the expression of FWA and FLOWERING LOCUS C and thus promote the floral transition. The Plant Cell, 19(10), pp.2975-2987.

Kantar, M., Unver, T. and Budak, H., 2010. Regulation of barley miRNAs upon dehydration stress correlated with target gene expression. Functional & integrative genomics, 10, pp.493-507.

Kazerooni, E.A., Maharachchikumbura, S.S., Adhikari, A., Al-Sadi, A.M., Kang, S.M., Kim, L.R. and Lee, I.J., 2021. Rhizospheric Bacillus amyloliquefaciens protects Capsicum annuum cv. Geumsugangsan from multiple abiotic stresses via multifarious plant growth-promoting attributes. Frontiers in plant science, 12, p.669693.

Khan, A.H., Min, L., Ma, Y., Zeeshan, M., Jin, S. and Zhang, X., 2023. High-temperature stress in crops: male sterility, yield loss and potential remedy approaches. Plant Biotechnology Journal, 21(4), pp.680-697.

Khraiwesh, B., Zhu, J.K. and Zhu, J., 2012. Role of miRNAs and siRNAs in biotic and abiotic stress responses of plants. Biochimica et Biophysica Acta (BBA)-Gene Regulatory Mechanisms, 1819(2), pp.137-148.

Kim, J.M., To, T.K., Nishioka, T. and Seki, M., 2010. Chromatin regulation functions in plant abiotic stress responses. Plant, Cell & Environment, 33(4), pp.604-611.

Kim, S.Y., Zhu, T. and Sung, Z.R., 2010. Epigenetic regulation of gene programs by EMF1 and EMF2 in Arabidopsis. Plant Physiology, 152(2), pp.516-528.

Kim, T.H., Böhmer, M., Hu, H., Nishimura, N. and Schroeder, J.I., 2010. Guard cell signal transduction network: advances in understanding abscisic acid, CO2, and Ca2+ signaling. Annual review of plant biology, 61, pp.561-591.

Kitsios, G. and Doonan, J.H., 2011. Cyclin dependent protein kinases and stress responses in plants. Plant signaling & behavior, 6(2), pp.204-209.

Knight, H., 1999. Calcium signaling during abiotic stress in plants. In International review of cytology (Vol. 195, pp. 269-324). Academic Press.

Kosová, K., Vítámvás, P., Urban, M.O., Prášil, I.T. and Renaut, J., 2018. Plant abiotic stress proteomics: the major factors determining alterations in cellular proteome. Frontiers in plant science, 9, p.122.

Kotak, S., Larkindale, J., Lee, U., von Koskull-Döring, P., Vierling, E. and Scharf, K.D., 2007. Complexity of the heat stress response in plants. Current opinion in plant biology, 10(3), pp.310-316.

Kotak, S., Larkindale, J., Lee, U., von Koskull-Döring, P., Vierling, E. and Scharf, K.D., 2007. Complexity of the heat stress response in plants. Current opinion in plant biology, 10(3), pp.310-316.

Kumar, S.V. and Wigge, P.A., 2010. H2A. Z-containing nucleosomes mediate the thermosensory response in Arabidopsis. Cell, 140(1), pp.136-147.

Kumar, V. and Wani, S.H. eds., 2020. Heat stress tolerance in plants: physiological, molecular and genetic perspectives. John Wiley & Sons.

Lämke, J. and Bäurle, I., 2017. Epigenetic and chromatin-based mechanisms in environmental stress adaptation and stress memory in plants. Genome biology, 18(1), pp.1-11.

Lämke, J., Brzezinka, K., Altmann, S. and Bäurle, I., 2016. A hit-and-run heat shock factor governs sustained histone methylation and transcriptional stress memory. The EMBO journal, 35(2), pp.162-175.

Larkindale, J. and Huang, B., 2004. Thermotolerance and antioxidant systems in Agrostis stolonifera: involvement of salicylic acid, abscisic acid, calcium, hydrogen peroxide, and ethylene. Journal of plant physiology, 161(4), pp.405-413.

Legris, M., Klose, C., Burgie, E.S., Rojas, C.C.R., Neme, M., Hiltbrunner, A., Wigge, P.A., Schäfer, E., Vierstra, R.D. and Casal, J.J., 2016. Phytochrome B integrates light and temperature signals in Arabidopsis. Science, 354(6314), pp.897-900.

Li, C., Ng, C.K.Y. and Fan, L.M., 2015. MYB transcription factors, active players in abiotic stress signaling. Environmental and Experimental Botany, 114, pp.80-91.

Li, L., Xue, M. and Yi, H., 2016. Uncovering microRNA-mediated response to SO2 stress in Arabidopsis thaliana by deep sequencing. Journal of hazardous materials, 316, pp.178-185.

Li, S., Liu, J., Liu, Z., Li, X., Wu, F. and He, Y., 2014. Heat-induced tas1 target1 mediates thermotolerance via heat stress transcription factor A1a–directed pathways in Arabidopsis. The Plant Cell, 26(4), pp.1764-1780.

Ling, Q., Broad, W., Trösch, R., Töpel, M., Demiral Sert, T., Lymperopoulos, P., Baldwin, A. and Jarvis, R.P., 2019. Ubiquitin-dependent chloroplast-associated protein degradation in plants. Science, 363(6429), p.eaav4467.

Liu, J.X. and Howell, S.H., 2010. bZIP28 and NF-Y transcription factors are activated by ER stress and assemble into a transcriptional complex to regulate stress response genes in Arabidopsis. The Plant Cell, 22(3), pp.782-796.

Liu, N.Y., Ko, S.S., Yeh, K.C. and Charng, Y.Y., 2006. Isolation and characterization of tomato Hsa32 encoding a novel heat-shock protein. Plant science, 170(5), pp.976-985.

Luo, M., Liu, X., Singh, P., Cui, Y., Zimmerli, L. and Wu, K., 2012. Chromatin modifications and remodeling in plant abiotic stress responses. Biochimica et Biophysica Acta (BBA)-Gene Regulatory Mechanisms, 1819(2), pp.129-136.

Manghwar, H. and Li, J., 2022. Endoplasmic reticulum stress and unfolded protein response signaling in plants. International Journal of Molecular Sciences, 23(2), p.828.

Mastrangelo, A.M., Marone, D., Laidò, G., De Leonardis, A.M. and De Vita, P., 2012. Alternative splicing: enhancing ability to cope with stress via transcriptome plasticity. Plant Science, 185, pp.40-49.

Mayer, M.P. and Bukau, B., 2005. Hsp70 chaperones: cellular functions and molecular mechanism. Cellular and molecular life sciences, 62, pp.670-684.

Meng, Y., Shao, C., Wang, H. and Jin, Y., 2012. Target mimics: an embedded layer of microRNA-involved gene regulatory networks in plants. BMC genomics, 13, pp.1-12.

Mishra, S.K., Tripp, J., Winkelhaus, S., Tschiersch, B., Theres, K., Nover, L. and Scharf, K.D., 2002. In the complex family of heat stress transcription factors, HsfA1 has a unique role as master regulator of thermotolerance in tomato. Genes & development, 16(12), pp.1555-1567.

Mittler, R., Finka, A. and Goloubinoff, P., 2012. How do plants feel the heat?. Trends in biochemical sciences, 37(3), pp.118-125.

Mittler, R., Vanderauwera, S., Gollery, M. and Van Breusegem, F., 2004. Reactive oxygen gene network of plants. Trends in plant science, 9(10), pp.490-498.

Moreno, A.A., Mukhtar, M.S., Blanco, F., Boatwright, J.L., Moreno, I., Jordan, M.R., Chen, Y., Brandizzi, F., Dong, X., Orellana, A. and Pajerowska-Mukhtar, K.M., 2012. IRE1/bZIP60-mediated unfolded protein response plays distinct roles in plant immunity and abiotic stress responses. PLoS One, 7(2), p.e31944.

Nakashima, K., Takasaki, H., Mizoi, J., Shinozaki, K. and Yamaguchi-Shinozaki, K., 2012. NAC transcription factors in plant abiotic stress responses. Biochimica et Biophysica Acta (BBA)-Gene Regulatory Mechanisms, 1819(2), pp.97-103.

Narsai, R., Law, S.R., Carrie, C., Xu, L. and Whelan, J., 2011. In-depth temporal transcriptome profiling reveals a crucial developmental switch with roles for RNA processing and organelle metabolism that are essential for germination in Arabidopsis. Plant physiology, 157(3), pp.1342-1362.

Noh, S.J. and Kwon, C.S., 2002. Characterization of two homologs of Ire1p, a kinase/endoribonuclease in yeast, in *Arabidopsis thaliana*. Biochimica et Biophysica Acta (BBA)-Gene Structure and Expression, 1575(1-3), pp.130-134.

Nover, L., Scharf, K.D., Gagliardi, D., Vergne, P., Czarnecka-Verner, E. and Gurley, W.B., 1996. The Hsf world: classification and properties of plant heat stress transcription factors. Cell stress & chaperones, 1(4), p.215.

Ohama, N., Sato, H., Shinozaki, K. and Yamaguchi-Shinozaki, K., 2017. Transcriptional regulatory network of plant heat stress response. Trends in plant science, 22(1), pp.53-65.

Palatnik, J.F., Allen, E., Wu, X., Schommer, C., Schwab, R., Carrington, J.C. and Weigel, D., 2003. Control of leaf morphogenesis by microRNAs. Nature, 425(6955), pp.257-263.

Pecinka, A., Rosa, M., Schikora, A., Berlinger, M., Hirt, H., Luschnig, C. and Scheid, O.M., 2009. Transgenerational stress memory is not a general response in Arabidopsis. PloS one, 4(4), p.e5202.

Pirayesh, N., Giridhar, M., Khedher, A.B., Vothknecht, U.C. and Chigri, F., 2021. Organellar calcium signaling in plants: An update. biochimica et biophysica Acta (bbA)-molecular Cell Research, 1868(4), p.118948.

Pratt, W.B., 1997. The role of Thehsp90-based chaperone system in signal transduction by nuclear receptors and receptors signaling via map kinase. Annual review of pharmacology and toxicology, 37(1), pp.297-326.

Probst, A.V., Fagard, M., Proux, F., Mourrain, P., Boutet, S., Earley, K., Lawrence, R.J., Pikaard, C.S., Murfett, J., Furner, I. and Vaucheret, H., 2004. Arabidopsis histone deacetylase HDA6 is required for maintenance of transcriptional gene silencing and determines nuclear organization of rDNA repeats. The Plant Cell, 16(4), pp.1021-1034.

Qiao, W. and Fan, L.M., 2008. Nitric oxide signaling in plant responses to abiotic stresses. Journal of Integrative Plant Biology, 50(10), pp.1238-1246.

Reis, R.S., 2023. Thermomorphogenesis: Opportunities and challenges in posttranscriptional regulation. Journal of experimental botany, p.erad134.

Richter, K., Haslbeck, M. and Buchner, J., 2010. The heat shock response: life on the verge of death. Molecular cell, 40(2), pp.253-266.

Rikhvanov, E.G., Fedoseeva, I.V., Pyatrikas, D.V., Borovskii, G.B. and Voinikov, V.K., 2014. Role of mitochondria in the operation of calcium signaling system in heat-stressed plants. Russian journal of plant physiology, 61, pp.141-153.

Ruberti, C., Kim, S.J., Stefano, G. and Brandizzi, F., 2015. Unfolded protein response in plants: one master, many questions. Current opinion in plant biology, 27, pp.59-66.

Saidi, Y., Finka, A. and Goloubinoff, P., 2011. Heat perception and signaling in plants: a tortuous path to thermotolerance. New phytologist, 190(3), pp.556-565.

Saidi, Y., Finka, A., Muriset, M., Bromberg, Z., Weiss, Y.G., Maathuis, F.J. and Goloubinoff, P., 2009. The heat shock response in moss plants is regulated by specific calcium-permeable channels in the plasma membrane. The Plant Cell, 21(9), pp.2829-2843.

Scharf, K.D., Berberich, T., Ebersberger, I. and Nover, L., 2012. The plant heat stress transcription factor (Hsf) family: structure, function and evolution. Biochimica et Biophysica Acta (BBA)-Gene Regulatory Mechanisms, 1819(2), pp.104-119.

Scharf, K.D., Berberich, T., Ebersberger, I. and Nover, L., 2012. The plant heat stress transcription factor (Hsf) family: structure, function and evolution. Biochimica et Biophysica Acta (BBA)-Gene Regulatory Mechanisms, 1819(2), pp.104-119.

Schröder, M. and Kaufman, R.J., 2005. ER stress and the unfolded protein response. Mutation Research/Fundamental and Molecular Mechanisms of Mutagenesis, 569(1-2), pp.29-63.

Schröder, M. and Kaufman, R.J., 2005. ER stress and the unfolded protein response. Mutation Research/Fundamental and Molecular Mechanisms of Mutagenesis, 569(1-2), pp.29-63.

Servet, C., e Silva, N.C. and Zhou, D.X., 2010. Histone acetyltransferase AtGCN5/HAG1 is a versatile regulator of developmental and inducible gene expression in Arabidopsis. Molecular plant, 3(4), pp.670-677.

Sharma P, Jha AB, Dubey RS, Pessarakli M. (2012). Reactive oxygen species, oxidative damage, and antioxidative defense mechanism in plants under stressful conditions. Journal of Botany, 2012, 217037. doi: 10.1155/2012/217037.

Sharma, S., Prasad, A., Sharma, N. and Prasad, M., 2021. Role of ubiquitination enzymes in abiotic environmental interactions with plants. International Journal of Biological Macromolecules, 181, pp.494-507.

Song, Z.T., Liu, J.X. and Han, J.J., 2021. Chromatin remodeling factors regulate environmental stress responses in plants. Journal of Integrative Plant Biology, 63(3), pp.438-450.

Sunkar, R., Li, Y.F. and Jagadeeswaran, G., 2012. Functions of microRNAs in plant stress responses. Trends in plant science, 17(4), pp.196-203.

Suzuki, N., Koussevitzky, S.H.A.I., Mittler, R.O.N. and Miller, G.A.D., 2012. ROS and redox signaling in the response of plants to abiotic stress. Plant, cell & environment, 35(2), pp.259-270.

Suzuki, N., Rivero, R.M., Shulaev, V., Blumwald, E. and Mittler, R., 2014. Abiotic and biotic stress combinations. New Phytologist, 203(1), pp.32-43.

Szabados, L. and Savouré, A., 2010. Proline: a multifunctional amino acid. Trends in plant science, 15(2), pp.89-97.

Szechyńska-Hebda, M., Ghalami, R.Z., Kamran, M., Van Breusegem, F. and Karpiński, S., 2022. To be or not to be? Are Reactive oxygen species, antioxidants, and stress signaling universal determinants of life or death?. Cells, 11(24), p.4105.

Tezuka, D., Matsuura, H., Saburi, W., Mori, H. and Imai, R., 2021. A ubiquitously expressed UDP-glucosyltransferase, UGT74J1, controls basal salicylic acid levels in rice. Plants, 10(9), p.1875.

Timperio, A.M., Egidi, M.G. and Zolla, L., 2008. Proteomics applied on plant abiotic stresses: role of heat shock proteins (HSP). Journal of proteomics, 71(4), pp.391-411.

Tiwari, Y.K. and Yadav, S.K., 2019. High temperature stress tolerance in maize (Zea mays L.): Physiological and molecular mechanisms. Journal of Plant Biology, 62, pp.93-102.

Tripathy, B.C. and Oelmüller, R., 2012. Reactive oxygen species generation and signaling in plants. Plant signaling & behavior, 7(12), pp.1621-1633.

Valmonte, G.R., Arthur, K., Higgins, C.M. and MacDiarmid, R.M., 2014. Calcium-dependent protein kinases in plants: evolution, expression and function. Plant and Cell Physiology, 55(3), pp.551-569.

Vera-Estrella, R., Barkla, B.J., Higgins, V.J. and Blumwald, E., 1994. Plant defense response to fungal pathogens (activation of host-plasma membrane H+-ATPase by elicitor-induced enzyme dephosphorylation). Plant Physiology, 104(1), pp.209-215.

Vierling, E., 1991. The roles of heat shock proteins in plants. Annual review of plant biology, 42(1), pp.579-620.

Wahid, A., Gelani, S., Ashraf, M. and Foolad, M.R., 2007. Heat tolerance in plants: an overview. Environmental and experimental botany, 61(3), pp.199-223.

Wang, G., Zeng, H., Hu, X., Zhu, Y., Chen, Y., Shen, C., Wang, H., Poovaiah, B.W. and Du, L., 2015. Identification and expression analyses of calmodulin-binding transcription activator genes in soybean. Plant and soil, 386, pp.205-221.

Wang, M., Vannozzi, A., Wang, G., Liang, Y.H., Tornielli, G.B., Zenoni, S., Cavallini, E., Pezzotti, M. and Cheng, Z.M.M., 2014. Genome and transcriptome analysis of the grapevine (*Vitis vinifera* L.) WRKY gene family. Horticulture research, 1.

Wang, M., Zou, Z., Li, Q., Xin, H., Zhu, X., Chen, X. and Li, X., 2017. Heterologous expression of three Camellia sinensis small heat shock protein genes confers temperature stress tolerance in yeast and *Arabidopsis thaliana*. Plant Cell Reports, 36, pp.1125-1135.

Wang, W., Vinocur, B., Shoseyov, O. and Altman, A., 2004. Role of plant heat-shock proteins and molecular chaperones in the abiotic stress response. Trends in plant science, 9(5), pp.244-252.

Wang, X., Cai, X., Xu, C., Wang, Q. and Dai, S., 2016. Drought-responsive mechanisms in plant leaves revealed by proteomics. International journal of molecular sciences, 17(10), p.1706.

Wang, X., Huang, W., Yang, Z., Liu, J. and Huang, B., 2016. Transcriptional regulation of heat shock proteins and ascorbate peroxidase by CtHsfA2b from African bermudagrass conferring heat tolerance in Arabidopsis. Scientific reports, 6(1), p.28021.

Wigge, P.A., 2013. Ambient temperature signaling in plants. Current opinion in plant biology, 16(5), pp.661-666.

Xiang, J., Ran, J., Zou, J., Zhou, X., Liu, A., Zhang, X., Peng, Y., Tang, N., Luo, G. and Chen, X., 2013. Heat shock factor OsHsfB2b negatively regulates drought and salt tolerance in rice. Plant cell reports, 32, pp.1795-1806.

Xu, P. and Yang, C., 2013. Emerging role of SUMOylation in plant development. Plant signaling & behavior, 8(7), p.e24727.

Zandalinas, S.I., Fritschi, F.B. and Mittler, R., 2020. Signal transduction networks during stress combination. Journal of Experimental Botany, 71(5), pp.1734-1741.

Zhang, S., Wang, S., Lv, J., Liu, Z., Wang, Y., Ma, N. and Meng, Q., 2018. SUMO E3 ligase SlSIZ1 facilitates heat tolerance in tomato. Plant and Cell Physiology, 59(1), pp.58-71.

Zheng, W., 2020. The plant sirtuins. Plant Science, 293, p.110434.

Zhou, S., Chen, Q., Sun, Y. and Li, Y., 2017. Histone H2B monoubiquitination regulates salt stress-induced microtubule depolymerization in Arabidopsis. Plant, cell & environment, 40(8), pp.1512-1530.

# CHAPTER 4

# SALICYLIC ACID (SA) AND NITRIC OXIDE (NO): THE PHENOMENAL PLAYERS IN PLANT HEAT STRESS RESPONSE


**Abstract**

Salicylic acid (SA) and nitric oxide (NO) are two major signaling molecules that regulate many growth and defence responses in plants exposed to various stress conditions. SA signaling has also been implicated in modulating HT stress response by regulating the expression of heat stress-related genes. SA-mediated NPR1 signaling activates heat shock proteins (HSPs) which act as molecular chaperones, protecting proteins from heat-induced denaturation. In addition, NO-signaling in plants occurs via the activity of nitric oxide synthase (NOS)-like enzymes or via regulating nitrate reduction. NOsignaling also activates the synthesis of HSPs and regulates plant defence response by modulating stomatal closure, thereby reducing water loss and preventing dehydration under heat stress. This chapter comprehensively explains the interactive role of SA and NO in plant heat stress. Further, the crosstalk of SA and NO with other phytohormones has also been described to unravel the signaling pathways involved in heat stress. Lastly, how the combined action of SA and NO help orchestrate a complex network of signaling events and physiological responses that enable plants to cope with heat stress is also critical.



**Keywords**: Salicylic acid, Nitric oxide, signaling, Heat, Crosstalk


## Introduction

SA is a naturally occurring compound found in plants. It is considered a plant hormone and plays a pivotal role in regulating plant growth, development, and defense against biotic and abiotic stress conditions (Vlot et al., 2009; Lefevere et al., 2020). Apart from regulating plant growth and development, SA also influences various other aspects such as seed germination, root development, flowering, and fruit ripening (Rivas-San Vicente and Plasencia, 2011). At the physiological level, SA also modulates stomatal closure, which affects water loss and gas exchange in plants (Kumar et al., 2012). Since its inception, SA has been considered a vital component of a plant's immune system, especially against pathogen attacks such as bacteria, fungi, or viruses (Lefevere et al., 2020). Increasing literature has well documented that exposure of plants to pathogens triggers the production and accumulation of salicylic acid in the infected tissues (Vlot et al., 2009). The increasing accumulation of SA helps to activate defense mechanisms by regulating the biosynthesis of antimicrobial compounds, cell wall integrity, and apoptosis in infected cells (Lefevere et al., 2020).

Later, further studies confronted the role of SA in stimulating plant's defense against abiotic stresses, such as drought, heat, cold, and heavy metal toxicity (Miura and Tada, 2014; Khan et al., 2015). They also intriguingly discovered that under stress conditions, SA acts as a signaling molecule, triggering the expression of defense-responsive genes (Horváth et al., 2007; Khan et al., 2015). Thus, efficiently modulating stress response and improving the plant's ability to withstand adverse environmental conditions (Dong et al., 2016). Interestingly, SA exerts the aforementioned defense response through a phenomenon called systemic acquired resistance (SAR), which is a long-lasting defense response that protects the entire plant (Durrant and Dong, 2004). Upon biotrophic attack, the transport of SA across different plant parts regulates defense-responsive genes/proteins, which enable them to defend against future attacks (Vlot et al., 2009; Horváth et al., 2007). In addition to its defense-responsive roles, SA, when applied exogenously, can trigger systemic resistance in plants. By doing so, exogenously used SA primes the plant's defenses and enables plants to withstand repeated episodes of biotic/abiotic stresses (Lefevere et al., 2020; Khan et al., 2015).

NO is a gaseous signaling molecule that is explicitly involved in the regulation of a wide array of plant functions and is involved in various physiological processes (Neill et al., 2002; Del Castello et al., 2019). Increasing literature has explained that NO functions as a signaling molecule in plants by activating communication between cells, thus coordinating different cellular and physiological processes (Yu et al., 2014; Arasimowicz and Floryszak-Wieczorek, 2007). NO also plays a significant role in improving plant functions under several biotic and abiotic stresses, such as pathogen attacks, drought, salinity, temperature extremes, and heavy metal toxicity (Bai et al., 2011; Khan et al., 2023;

Arasimowicz et al., 2007). Several reports have documented NO-mediated improvement in plant growth and development processes, particularly seed germination, root development, shoot elongation, flowering, and senescence (Arasimowicz and Floryszak-Wieczorek, 2007; Neill et al., 2003).

Furthermore, under stress conditions, NO also regulates stomatal movement, which controls gas exchange and water loss in plants after interacting with other plant hormones, such as auxins, gibberellins, and abscisic acid (Neill et al., 2002; Isner and Maathuis, 2016; Khan et al., 2018). NO-mediated production of other signaling molecules has been shown to promote stomatal opening by activating ion channels, thereby preventing the adverse effects of stress-induced oxidative damage in plants (García-Mata and Lamattina, 2001; Gao et al., 2017). NO is also actively involved in combating biotic and abiotic stresses (Zhang et al., 2022). NO acts as a signaling molecule to activate defense signaling pathways, thus modulating the expression of several defense-responsive genes (Domingos et al., 2015). This transcriptional activation of genes, in turn, activates the biosynthesis of other signaling compounds, thus improving plant growth and productivity under stress conditions (Asgher et al., 2017). NO also functions by interacting with other signaling molecules such as salicylic acid, jasmonic acid, and abscisic acid (Neill et al., 2002; Arasimowicz and Floryszak-Wieczorek, 2007). These interactions form complex signaling networks that regulate plant responses to various stimuli and coordinate multiple physiological processes (Yu et al., 2014). Both SA and NO are versatile signaling molecules capable of regulating plant growth, development, and stress responses (Table 1; Figure 1). Their precise functions and mechanisms of action are still being studied, and ongoing research continues to uncover new insights into the role of SA and NO in plant biology.

## Salicylic acid biosynthesis in plants

SA is a phytohormone that plays a notable role in regulating various physiological and cellular processes, including stress response, thus improving plant growth and development (Miura and Tada, 2014). The biosynthesis of SA in plants occurs through multiple pathways; however, the most well-known biosynthetic pathway is the isochorismate synthase (ICS) pathway. This pathway is widely conserved among higher plants and is the primary route for salicylic acid production. The first step in SA biosynthesis is the conversion of the central metabolite, 5-

enolpyruvylshikimate-3-phosphate (EPSP), to chorismate via chorismate synthase (Mauch-Mani and Slusarenko, 1996). The enzyme isochorismate synthase (ICS) catalyzes the isomeric conversion of chorismate to isochorismate, which is then used as a precursor for SA synthesis in plants (Wildermuth et al., 2001; Garcion et al., 2008). The isochorismate pyruvate lyase (IPL) and salicylate synthase (SAS) are explicitly involved in the biosynthetic process. The enzyme IPL decarboxylates isochorismate to produce pyruvate and salicylate. The generated salicylate is then converted into salicylic acid by the enzyme salicylate synthase (Lefevere et al., 2020; Vlot et al., 2019).

Another major route for the biosynthesis of SA in plants is the phenylpropanoid pathway, which involves the conversion of a phenylalanine-derived precursor called cinnamic acid to SA (Vlot et al., 2009; Maeda and Dudareva, 2012). The biosynthetic pathway begins with the enzyme Phenylalanine ammonia-lyase (PAL), which converts the amino acid phenylalanine into cinnamic acid (Dixon et al., 2002). Next, the cinnamic acid is hydroxylated to form 4-coumaric acid in a reaction catalyzed by the enzyme Cinnamate 4-hydroxylase (C4H). The enzyme 4-Coumarate CoA ligase activates 4-coumaric acid, which is then converted into 4-coumaroyl-CoA (Fraser and Chapple, 2011). The activated acid is linked to coenzyme A (CoA) for further modifications in the above step.

4-coumaroyl-CoA is then reduced to cinnamoyl-CoA by the enzyme Cinnamoyl-CoA reductase. Further, Cinnamoyl-CoA is then benzoylated by the enzyme Benzyl alcohol O-benzoyltransferase, producing benzoyl-CoA (Widhalm et al., 2015). This Benzoyl-CoA is hydrolyzed by the enzyme Salicylic acid biosynthetic protein, releasing benzoyl and CoA. The free benzoyl group is subsequently converted to SA via a non-enzymatic decarboxylation reaction (Widhalm et al., 2015).

However, The ICS pathway is considered the most significant pathway for salicylic acid production in response to biotic and abiotic stress conditions. The biosynthesis of SA in plants is initiated upon the inception of defence-related signaling pathways, such as the plant immune response mediated by the signaling molecule known as salicylic acid. Correspondingly, it's worth mentioning the existence of other biosynthetic pathways for salicylic acid in plants which may be more prevalent in certain plant species or under specific physiological conditions. Nonetheless, the ICS pathway is the most widely recognised and extensively studied for plant salicylic acid biosynthesis.

**Table 1** Role of salicylic acid (SA) and nitric oxide (NO) signaling in influencing growth and productivity in different plants under heat stress

| Plant Species | Role of Salicylic Acid (SA) | Role of Nitric Oxide (NO) | References |
|---|---|---|---|
| Tomato | SA enhances heat tolerance by regulating antioxidant systems, mitigating oxidative damage, and maintaining photosynthetic efficiency. | NO acts as a signaling molecule to induce heat stress-responsive genes, enhance antioxidant capacity, and promote thermotolerance. | Neill et al. (2002) Baudouin, E. and Hancock, J.T. (2014) |
| Wheat | SA improves heat tolerance by enhancing antioxidant defense mechanisms, reducing oxidative stress, and improving photosynthetic performance. | NO alleviates heat-induced damage by regulating antioxidant enzymes, reducing oxidative stress, and maintaining membrane stability. | Li et al. (2013); Du et al. (2012) |
| Maize | SA enhances heat tolerance by regulating the expression of heat shock proteins (HSPs), activating antioxidant enzymes, and improving photosynthetic efficiency. | NO modulates heat stress responses by regulating HSP expression, activating antioxidant defense systems, and improving photosynthetic capacity. | Ding et al. (2018); Zhang et al. (2018a) |
| Soybean | SA confers heat tolerance by modulating antioxidant defense systems, regulating stomatal closure, and improving photosynthetic performance. | NO acts as a signaling molecule in heat stress responses, promoting antioxidant activity, enhancing photosynthetic efficiency, and improving yield under heat stress. | Zhang et al. (2018a); Baudouin, E. and Hancock, J.T. (2014) |

| | | | |
|---|---|---|---|
| Rice | SA enhances heat tolerance by regulating the expression of heat-responsive genes, activating antioxidant enzymes, and improving photosynthetic efficiency. | NO plays a crucial role in heat stress tolerance by enhancing antioxidant defense systems, regulating stomatal conductance, and improving photosynthetic capacity. | Srivastava et al. (2014); Yu et al. (2019) |

## Nitric oxide biosynthesis in plants

NO is a highly reactive and short-lived gaseous signaling molecule that is exceptionally involved in regulating vital biological processes in plants (Neill et al., 2002; Astier et al., 2011). NO is also known to stimulate signaling cascades, thus regulating plant growth, development, and immune responses to environmental stresses (Yu et al., 2014). Nitric oxide is synthesized in plants through enzymatic and non-enzymatic pathways. The enzymatic pathway involves the Nitrate Reductase (NR) pathway. In this pathway, nitrate reductase produces NO from nitrate ($NO_3^-$). Nitrate reductase reduces nitrate to nitrite ($NO_2^-$) and generates NO as a byproduct (Rockel et al., 2002). Another class of enzyme, Nitric Oxide Synthase-Like (NOSL) enzymes, is also involved in the biosynthesis of NO in plants (Lozano-Juste and León, 2011). The NOSL found in plants are analogous to animal nitric oxide synthases (NOS); however, the exact mechanism by which the NOSL pathway plays a role in NO biosynthesis is still unclear.

Another enzyme that catalyzes NO biosynthesis in plants is Xanthine Oxidoreductase (XOR), which also exclusively regulates purine metabolism in plants. Several studies have suggested that XOR has NOS-like activity in plants, which can produce NO through the oxidation of xanthine or hypoxanthine in the presence of molecular oxygen (Corpas et al., 2019; Yu et al., 2014). Likewise, S-Nitrosoglutathione Reductase (GSNOR) is an enzyme documented to play a critical role in NO biosynthesis in plants (Feechan et al., 2005). The GSNOR mechanism stimulates NO biosynthesis by performing the catalytic breakdown of S-nitrosoglutathione (GSNO). This catalytic degradation of GSNO triggers NO release and modulates intracellular NO levels. Recently, researchers have also pinpointed the role of non-symbiotic hemoglobins in NO biosynthesis (Feechan et al., 2005). Few plant hemoglobins, such as class I and II non-symbiotic hemoglobins, are actively involved in the scavenging of NO radicals, thus regulating its bioavailability.

Later findings have also implicated the involvement of a non-enzymatic pathway of nitric oxide (NO) biosynthesis in plants through a series of chemical reactions and without the direct participation of enzymes (Beligni and Lamattina, 2000). One such pathway involves a nitrite-dependent way that triggers $NO_2^-$ reductions via ascorbic acid or reduced glutathione under specific conditions. This spontaneous reduction of $NO_2^-$ contributes to NO production under low-oxygen conditions (Neill et al., 2002). The reactive nitrogen species peroxynitrite ($ONOO^-$), which is formed by the reaction between NO and superoxide anion, has also been reported to stimulate NO production in plants (Yun et al., 2011). Peroxynitrite can spontaneously react with enzymatic and non-enzymatic antioxidants, resulting in the formation of NO in plants (Wu et al., 2011). Similarly, certain transition metals such as Cu and Fe can stimulate nitric oxide production in plants by a process called metal-catalyzed NO generation. These transition metals catalyze the transfer of electrons and stimulate the reduction of nitrite to NO (Bethke et al., 2007).

Despite the above understanding, the mechanistic insight into enzymatic and non-enzymatic pathways for NO biosynthesis in plants is yet to be explored. Researchers have reported the existence of some enzymes in plants that regulate NO synthesis in plants. However, the presence and role of a genuine NOS enzyme in plants still seem far-fetched and require extensive investigation. Conversely, the non-enzymatic pathways of NO biosynthesis in plants are regulated by many cellular and physiological factors, such as pH, redox conditions, reactive oxygen species (ROS), and the availability of substrates.

## Salicylic acid signaling in plant heat stress

Heat stress is a ferocious abiotic condition that savagely affects plant growth and development (Table 2). Under stress conditions, SA signaling has been known to unequivocally improve various cellular and physiological processes in plants, including growth and productivity (Shivaraj et al., 2020). Earlier studies have restricted the involvement of SA to providing defense against insects and pathogens (Vlot et al., 2009). However, emerging studies suggest that SA can regulate plant growth, development, and productivity under abiotic stress conditions. Recent findings indicate that heat stress triggers the accumulation of SA in plant tissues by stimulating the production of reactive oxygen species (ROS).

These ROS serve as signaling molecules when their levels are maintained below a certain threshold (Savvides et al., 2016). Beyond the threshold, the generation of ROS starts instigating oxidative damage in plants, leading to a dynamic increase in the level of SA that stimulates ROS degradation (Mhamdi et al., 2012). SA signaling in plants also functions through Nonexpressor of PR genes 1 (NPR1), which is a crucial component in the activation of SA-mediated defense signaling in plants (Spoel et al., 2003).

NPR1 functions as a central regulator of SA signaling in plants exposed to climate extremes. An increasing body of literature has documented NPR1 as a critical component of SA-mediated defense responses. NPR1 activates SA-mediated defense responses by coordinating the transcriptional activation of defense genes and regulating innate immune responses (Wu et al., 2012). Exposure of plants to biotic and abiotic stresses gradually accelerates SA production, which leads to the activation of NPR1 (Molinier et al., 2006). Under normal conditions, the oligomeric form of NPR1 is present in the cytoplasm. However, upon stress exposure, the binding of SA to NPR1 induces a conformational change in its structure, which is an essential prerequisite for its subsequent monomerization and translocation into the nucleus (Spoel et al., 2003). Within the nucleus, NPR1 interacts with TGA transcription factors, leading to their activation. The activated NPR1-TGA complex recognizes and binds to specific DNA sequences known as cis-regulatory elements, thereby activating the transcription of defense-related proteins such as pathogenesis-related 1 (PR1) and other SA-responsive genes (Cao et al., 1997).

In addition, studies have also corroborated that NPR1 acts as a co-activator by recruiting various co-regulators to modulate gene expression. Interaction of NPR1 with co-regulators, such as NIMINs (NPR1-interacting proteins) and histone acetyltransferases (HATs), modulates chromatin remodeling and transcriptional activation of target genes (Spoel et al., 2007; Van Verk et al., 2009). Furthermore, NPR1-mediated SA accumulation and subsequent interaction with transcription factors and co-regulators develop systemic acquired resistance (SAR). SAR is a systemic defense response that is created in the distal parts of the plant following pathogen infection/oxidative damage (Durrant and Dong, 2004). Further, the SAR pathway also interacts with and activates other signaling pathways, such as jasmonic acid and abscisic acid pathways, to alleviate stress induces oxidative damage in plants by activating defence gene expression in systemic tissues, priming them for enhanced resistance to subsequent stress exposure.

**Table 2** Salicylic acid (SA) and nitric oxide (NO) in influencing various physiological and molecular processes in different plants under stress conditions

| Physiological/Molecular Process | Salicylic Acid (SA) | Nitric Oxide (NO) | References |
|---|---|---|---|
| Stomatal regulation | SA can induce stomatal closure, reducing water loss and preventing pathogen entry. | NO regulates stomatal aperture, promoting stomatal closure to reduce water loss and prevent pathogen invasion. | Liu et al. (2009) |
| Photosynthesis | SA enhances photosynthesis by modulating the expression of genes involved in photosynthetic machinery and protecting photosystem II from oxidative damage. | NO regulates photosynthesis by modulating the activity of key enzymes involved in carbon fixation and photosystem II repair. | Liu et al. (2009a) |
| Antioxidant defense | SA activates antioxidant enzymes, such as superoxide dismutase (SOD) and catalase, to scavenge reactive oxygen species (ROS) and mitigate oxidative stress. | NO acts as a signaling molecule to induce the expression of antioxidant enzymes, including SOD, catalase, and peroxidase, thereby reducing oxidative damage. | Liu et al. (2019) |
| Pathogen defense | SA triggers systemic acquired resistance (SAR), activating defense genes and enhancing plant resistance against pathogens. | NO acts as a signal molecule in plant defense responses, activating defense genes and enhancing pathogen resistance. | Prakash et al. (2021) |

| | | | |
|---|---|---|---|
| Regulation of gene expression | SA modulates the expression of stress-responsive genes, including transcription factors and pathogenesis-related (PR) genes. | NO regulates gene expression by modifying chromatin structure and modulating the activity of transcription factors involved in stress responses. | Luna et al. (2012) |
| Root development | SA promotes lateral root formation and enhances root growth under stress conditions. | NO regulates root development by modulating root architecture and promoting lateral root formation. | Luna et al. (2012) |
| Abscisic acid (ABA) signaling | SA interacts with ABA signaling pathways, influencing stomatal closure and regulating stress responses. | NO interacts with ABA signaling pathways, enhancing ABA-induced stomatal closure and stress tolerance. | Prakash et al. (2021) |

## Nitric oxide signaling in plant heat stress

NO signaling in plants is regulated by Nitric Oxide Synthase (NOS)-like enzymes which are analogous to animal NOS enzymes, although their structure and regulation may differ (Guo et al., 2003). However, in plants, the existence of NOS-like enzymes similar to those found in animals has been a subject of intense debate (Corpas et al., 2019). Recent findings have implicated that NO signaling under stress conditions is mediated by the genes encoding NOS-like enzymes that are involved in NO synthesis in plants (Rockel et al., 2002). Another aspect of NO-mediated signaling in plants involves GSNOR (S-nitroso glutathione reductase) enzyme (Feechan et al., 2005). GSNOR acts as a critical component of modulating NO homeostasis by effectively regulating S-nitroso glutathione (GSNO) levels (Liu et al., 2001). Further, GSNOR also substantially acts as a major NO transporter and reservoir in plants (Liu et al., 2001).

GSNOR is also actively involved in the degradation of GSNO, and other specific conditions release NO, thus acting as a source of bioactive NO (Liu et al., 2001). In addition, GSNOR also regulates and maintains GSNO levels by converting it into oxidized glutathione (GSSG) and ammonia, thus effectively regulating the bioavailability of NO (Feechan et al., 2005). A large body of literature has also confirmed the influential role of GSNOR in regulating various NO-dependent physiological processes in plants (Yun et al., 2011). Upon stress exposure, the release of NO from GSNO instantly modifies the target proteins through S-nitrosylation, a post-translational modification involving the transfer of an NO group to a cysteine residue in proteins (Yun et al., 2011). S-nitrosylation can modulate the innate immune response of plants by regulating protein activity, stability, and interactions, thus activating downstream stress signaling genes (Astier et al., 2011).

Several studies have also confirmed that GSNOR-mediated NO signaling influences various other aspects of plant development, such as seed germination, root development, flower development, and senescence (Zhou et al., 2005; Bhattacharjee and Bhattacharjee, 2019). S-nitrosylation of specific stress-responsive genes/proteins is known to widely regulate the above-mentioned developmental processes by activating gene expression and signaling pathways (Lozano-Juste and Leon, 2011; Terrile et al., 2012). Concomitantly, GSNOR-mediated NO signaling under heat stress also functions by interacting with phytohormone signaling pathways in plants. NO and GSNO have been documented to interact with hormones such as jasmonic acid, auxins, ethylene, abscisic acid (ABA), and gibberellins (GAs), influencing heat stress tolerance in plants (Freschi, 2013; Devireddy et al., 2021; Shivaraj et al., 2020). GSNOR-mediated NO signaling is an imperative mechanism that regulates the bioavailability of NO in plants. The increased NO production modulates many physiological processes, stress defence responses, and developmental processes. However, future investigation must unravel the specific targets and mechanisms underlying GSNOR-mediated NO signaling in different plant species.

## Role of salicylic acid in mitigating heat stress

The potential of SA in mitigating the adverse effect of heat stress in plants has long been studied, and several findings have confirmed the possibility of SA in enhancing heat stress tolerance in plants (Rivas-San Vicente and Plasencia, 2011). SA helps plants perceive heat stress, thus activating defense signaling pathways (Table 2). Once started, these defense signaling triggers several physiological and biochemical responses which protect the plants from oxidative damage (Nabi et al., 2019; Rai et al.,

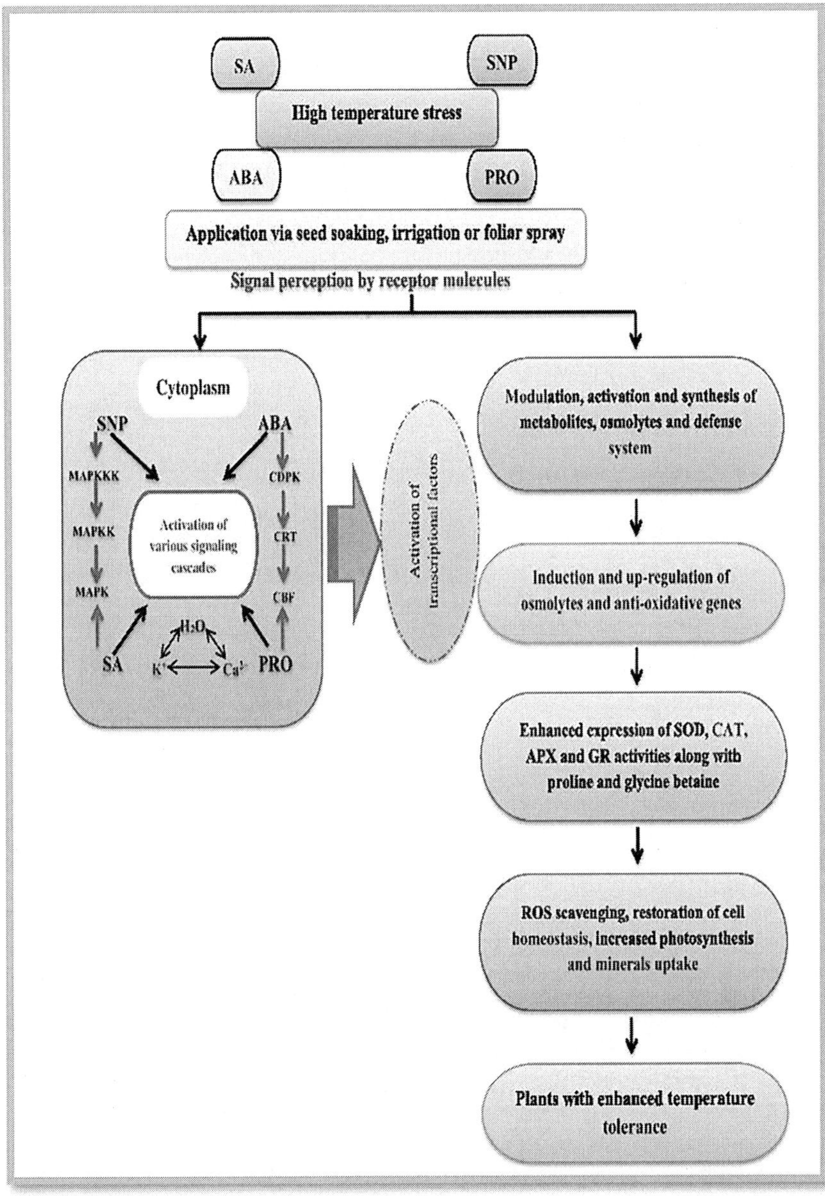


**Figure 1** Schematic representation of phytohormone mediated stimulation of heat stress tolerance in plants

2020). SA decapitates heat stress in plants by activating the enzymatic and non-enzymatic antioxidants that effectively regulate ROS-mediated oxidative damage. In addition, SA also improves photosynthetic capacity and membrane integrity by effectively maintaining redox balance and carbon assimilation under heat stress (Rai et al., 2020). SA also exerts a protective response by modulating post-translational modification, transcription factors, and epigenetic modification in plants exposed to HT stress (Luo et al., 2015; Xu et al., 2017). Several studies have reported that the exogenous application of SA in mitigating plant stress-induced oxidative damage is practical. Still, the tolerance level varies depending upon species, age, and type of application (AbdElgawad et al., 2016; Gill et al., 2013).

## Transcription factor-mediated regulation

An increasing body of literature has delineated that SA regulates the growth and development of plants exposed to climate extremes by modulating the signaling pathway primarily involved in the activation of various transcription factors (Miura and Tada, 2014; Xuan et al., 2010). Transcription factors are the regulatory proteins that regulate the transcription of different downstream signaling genes after binding to specific DNA sequences (Zhang et al., 2006 Dong et al., 2019). For example, the WRKY family transcription factors are widely known to stimulate defense responses in plants exposed to biotic/abiotic stress by regulating salicylic acid signaling (Birkenbihl et al., 2017; Rai et al., 2021). Several findings have reported that stress-induced activation of SA signaling modulates WRKY transcription factors after binding to specific DNA sequences known as W-boxes and regulates the expression of defense-related genes (Li et al., 2021). Once activated, these WRKY TFs activate the biosynthesis and expression of various defense-related proteins, such as PR proteins and enzymes involved in plant stress tolerance (Rai et al., 2021).

NPR1 also acts as a transcriptional activator of SA signaling, which then acts as a transcription co-activator that interacts and modulates the activity of other transcription factors (Molinier et al., 2006; Khurana et al., 2017). SA also regulates the translocation of NPR1 into the nucleus, where in conjunction with WRKY transcription factors, it initiates the transcription of defense-responsive genes (Spoel et al., 2003; Fu et al., 2012). In addition, the translocated NPR1 also interacts with other transcription factors, such as TGA factors, and activates their transcription to improve plant defense response (Xu et al., 2017). TGA transcription factors have also been known to regulate a plethora of physiological and biochemical processes in plants, including stress responses (Luo et al., 2020; Fonseca et al., 2022). SA-mediated signaling triggers the activation of transcription of TGA in the nucleus, where it interacts with NPR1 and WRKY transcription factors to regulate the expression of defense genes (Luo et al., 2020). TGA factors exert a defense response after binding to specific DNA sequences known as as-1 elements, which are present in the promoters of target genes (Bai et al., 2011; Johnson et al., 2003). The SA-mediated binding of TGA TFs with these elements contributes to the activation of plant innate immunity and strengthens the plant's defense against environmental cues (Bai et al., 2011; Johnson et al., 2003).

MYB and MYC transcription factors have also been documented to play a unique role in regulating different physiological processes in plants, particularly under stress conditions (Pauwels et al., 2010). Further, salicylic acid signaling-mediated NPR1 activation and subsequent translocation into the nucleus activate the expression of MYB TFs (Ushimaru et al., 2006; Bai et al., 2011). MYB TFs are profoundly involved in the regulation of flavonoid biosynthesis in plants, which contributes explicitly to plant defense against UV-B and pathogens (Gao et al., 2017; Mao et al., 2020). Similarly, SA signaling also influences the transcription of MYC transcription factors, which are explicitly involved in mediating jasmonic acid and ethylene biosynthesis to regulate plant defense responses (Lorenzo et al., 2004; Kazan and Manners, 2013).

## Epigenetic regulation

SA has been found to play a role in epigenetic regulation in plants that regulate gene expression that occur without alterations in the DNA sequence (Zhang et al., 2006). Under stress conditions, SA signaling regulates DNA methylation in plants which involves the addition of a methyl group to cytosine residues in DNA, often occurring at CG sites (Liu et al., 2004). Further, SA also induces DNA hypomethylation, thus differentially regulating the expression of various defence-responsive genes/proteins (Hu et al., 2017). For example, a plethora of researchers have corroborated that external applications of SA have been shown to improve plant growth and development under heat stress by activating hypomethylation leading to increased expression of defence-related genes (Khan et al., 2015). SA-induced DNA demethylation has been observed in different plant species, including Arabidopsis, tobacco, and rice, which has improved their tolerance under stress conditions by activating defence-related genes (Rubio-Somoza and Weigel, 2011). The mechanism by which SA mediates DNA methylation in plants is the enzymatic inhibition of DNA methyltransferases (DNMTs) enzymes which catalyze the addition of methyl groups to DNA (Mao et al., 2020). Another mechanism is the SA-mediated activation of DNA demethylase enzymes, which is potentially involved in the removal of methyl groups from DNA (Chen et al., 2010). SA treatment has been linked to the increased expression of genes encoding DNA demethylases, potentially contributing to DNA demethylation (Spoel et al., 2003).

Salicylic acid also influences histone modifications in plants exposed to biotic/abiotic stress conditions, such as acetylation and methylation, causing conformational changes in chromatin structure and thus regulating gene expression patterns (Rai et al., 2020). Foliar application of SA has been proclaimed to improve histone acetylation, thus activating transcription of various stress-responsive genes in plants under HT stress (Pandey et al., 2016). Furthermore, alterations in histone methylation patterns have also been shown to downregulate gene expression states, thus affecting stress defence response in plants (Berr et al., 2011). Several studies have delineated that SA differentially regulates the activity of histone deacetylase (HDAC) enzyme, which catalyzes the removal of acetyl groups from histones (Van Verk et al., 2018). The SA-mediated de-acetylation of histones provokes chromatin condensation and transcriptional repression (Yu et al., 2009). SA-mediated regulation of HDACs induces a conformational change in the chromatin structure, thus activating the transcription of stress-responsive genes/proteins (Prakash et al., 2021). Studies have also indicated the mechanistic intersection of SA's mediated histone modifications t with other epigenetic marks, such as DNA methylation. Several findings have observed that SA-mediated DNA methylation interacts with histone modifications to activate gene expression. The interplay between SA, DNA methylation, and histone modifications contributes to the complex epigenetic regulation of gene expression in response to various stimuli and stresses.

## Role of nitric oxide in mitigating heat stress

NO is a gaseous reactive molecule that plays distinct regulatory roles in plants as well as in mitigating heat stress (Gautam et al., 2021). Reports have confirmed it as a key player in enhancing stress tolerance in plants by stimulating stress signaling pathways as well as by triggering a myriad of signaling cascades to render HT tolerance in plants (Nabi et al., 2019). NO does this by contriving the generation of ROS beyond a certain threshold, thus effectively modulating them to function as signaling molecules (Fan et al., 2013). It also stimulates various enzymatic and non-enzymatic antioxidants such as CAT, SOD, GR, and APX that effectively minimize the ROS-induced oxidative damage in plants (Wu et al., 2011). Further studies have also reported that NO actively regulates the transcription of transcription factors in plants, thus regulating gene expression in response to different biotic/abiotic stresses (Khurana et al., 2017).

## Transcription factor-mediated regulation

In plants, one of the most extensively studied NO-mediated activations of TFs is Hypoxia-Inducible Factor 1 (HIF-1), a TF which is exclusively activated in response to low-oxygen conditions (hypoxia) (Gupta and Igamberdiev, 2016). Various reports have implicated NO to regulate and stabilize HIF-1 by modulating the activity of prolyl hydroxylase enzymes, which degrade the HIF-1 target under normoxic (normal oxygen) conditions (Bai et al., 2011). NO-mediated stabilization and translocation of HIF-1 to the nucleus activate the expression of genes involved in hypoxia adaptation in plants (Chen et al., 2010). Similarly, NO signaling is also implicated in modulating the expression of HSFs under heat stress conditions (Zhang et al., 2006). Several studies have confronted the role of NO in regulating the effective binding of HSFs to the corresponding DNA element, which enhances the expression of HSF genes (Christou et al., 2014). The upregulated HSF genes then activate the synthesis of HSPs that help protect plants from heat-induced damage (Xu et al., 2017).

Another class of TFs are ethylene-responsive factors (ERF) which regulate plant growth, development, and stress responses, including the biosynthesis of the hormone ethylene (Rubio-Somoza and Weigel, 2011). Several recent findings have confirmed the regulatory role of NO in modulating the activity of ERFs (Wang et al., 2002). In several plant species, NO signaling has upregulated the DNA binding activity of ERFs, leading to the enhanced expression of ethylene-responsive genes (Xu et al., 2017). Recent findings have also corroborated that the interaction between NO and ERFs can be complex, as NO-mediated post-translational modifications may lead to differential regulation of ERFs, i.e., can have both activating and inhibitory effects on their function (Manjunatha et al., 2010).

Another class of TFs which are known to be modulated by NO signaling is the Nuclear Factor Y (NF-Y), which also plays a significant role in diverse biological processes, including activation of stress defense responses

(Besson-Bard et al., 2008). Exogenous application of NO has been shown to regulate the activity of NF-Y YFs in several plant species (Parankusam et al., 2017). In Arabidopsis, foliar application of NO regulated the binding of NF-Y for target gene promoters, thus enhancing their transcription and expression (Zhang et al., 2006). This NO-mediated modulation of NF-Y contributes to the regulation of stress-responsive genes, which strengthens plant growth and defense response under HT stress (Rai et al., 2020).

Exogenous applications of NO have also been known to regulate the expression of several stress-specific transcription factors, such as DREB2A (dehydration-responsive element-binding protein 2A). DREB TFs are widely acclaimed stress-responsive TFs which are explicitly involved in drought, salt, and cold stress responses (Liu et al., 2022). Extensive studies have been conducted to decipher the role of NO in stimulating the accumulation of DREB2A mRNA in plants exposed to HT stress (Rai et al., 2020). For example, in rice and Arabidopsis, foliar application of NO activated the transcription of DREB TFs, which in turn started the expression of several stress-responsive genes/proteins involved in stress tolerance (Rai et al., 2020).

Likewise, NO signaling also stimulates abiotic stress tolerance by modulating the expression of Abscisic Acid (ABA)-Responsive Transcription Factors. ABA is a well-known phytohormone that regulates various physiological processes in plants under different stress conditions (Fujita et al., 2004). An increasing body of literature has restricted the involvement of NO signaling in modulating the activity of ABA-responsive transcription factors in several plant species (Zhang). For instance, the external application of NO positively regulated the binding of ABRE-binding factors (ABFs), which led to enhanced expression of ABA-responsive gene expression in Arabidopsis plants exposed to HT stress conditions (Zhang et al., 2006).

## Epigenetic regulation

NO, the gaseous signaling molecule, has emerged as a significant player in modulating plant growth and defense responses under various biotic and abiotic stresses (Neill et al., 2002). Despite its profound role at the physiological and biochemical level, NO can also mediate stress response at the molecular level by regulating epigenetic modifications, thus affecting gene expression and plant adaptive responses. One such change involves DNA methylation, which negatively regulates the expression of various genes/proteins in plants exposed to stress conditions (Yu et al., 2011). Nonetheless, exogenous application of NO has been shown to revert the DNA methylation pattern, i.e., promote DNA demethylation that enhances the expression of various defense-responsive genes (Liu et al., 2004). Several investigations have revealed that NO either directly or indirectly modulates DNA methylation patterns by regulating the activity of DNA methyltransferases (DNMTs) activity. However, the exact mechanism by which NO regulates the activity of DNMTs varies greatly depending on the stress conditions and plant species (Christou et al., 2014). Nonetheless, researchers have revealed that it may be due to NO-mediated inhibition of DNMT activity, leading to global DNA demethylation. Conversely, in some cases, NO has also been found to induce DNMT activity, resulting in increased DNA methylation of specific genes leading to their decreased expression (Liu et al., 2004; Yu et al., 2011).

GSNOR is an enzyme that catalyzes the degradation of GSNO, thus regulating NO biosynthesis and its signaling activity (Feechan et al., 2005). Several reports have also confirmed that GSNOR-mediated NO signaling can influence DNA methylation patterns in plants exposed to biotic/abiotic stresses (Zhou et al., 2005). For instance, in Arabidopsis thaliana, GSNOR stimulates DNA methylation of several stress-responsive genes and alters their expression, leading to decreased growth under stress conditions (Zhou et al., 2005). NO signaling has also been implicated in enhancing the expression of demethylases, the enzymes known to remove methyl groups from DNA, thereby improving DNA methylation marks. In Arabidopsis, the external application of NO stimulated the expression of the demethylase ROS1 (Repressor of Silencing 1), which is actively involved in the removal of DNA methylation from specific genomic regions. The removal of DNA methylation marks activated the expression of DREB and HSP, improving plant growth and immunity under drought stress.

Several Small RNAs, such as small interfering RNAs (siRNAs) and microRNAs (miRNAs), are known to differentially regulate biotic/abiotic stress tolerance in plants (Katiyar et al., 2017; Khraiwesh et al., 2012). These small RNAs exert this regulatory function by activating post-transcriptional regulation (Katiyar et al., 2017; Khraiwesh et al., 2012). Researchers have indicated that NO signaling also regulates DNA methylation by modulating the RNA-directed DNA methylation (RdDM) pathways (Liu et al., 2004; Meng et al., 2011). In addition, NO signaling also modulates the biogenesis of other small RNA genes, such as DICER-

LIKE (DCL) and ARGONAUTE (AGO) family genes, which also regulate DNA methylation patterns in plants under heat stress (Liu et al., 2004; Meng et al., 2011).

## SA and NO-mediated post-translation modification in plant heat stress

SA signaling in plants can trigger post-translational modifications (PTMs) of several stress-responsive proteins under stress conditions (Liu et al., 2016; Rivas-San Vicente et al., 2011). SA signaling stimulates plant stress defense responses by modulating protein phosphorylation cascades (Feki et al., 2021; Mishina & Zeier, 2007). SA activates phosphorylation cascades by influencing the activity of protein kinases, the enzymes that catalyze the transfer of phosphate groups to target proteins, thereby stimulating post-translational modifications (Durrant & Dong, 2004; Mishina & Zeier, 2007). This phosphorylation event upregulates the activity, stability, and localization of stress-responsive proteins and can alter their function, ultimately improving plant defense responses (Liu et al., 2016; Feki et al., 2021). Acetylation is a PTM that involves the addition of an acetyl group to lysine residues of proteins (Belkhadir et al., 2004; Ruan et al., 2001). SA-mediated acetylation has been found to play a regulatory role in plant growth and defense (Liu et al., 2016; Belkhadir et al., 2004). The enzyme that mediates SA-mediated acetylation in plants is salicylic acid-binding protein 2 (SABP2), which activates downstream signaling events. SABP2-mediated SA-acetylation activates and regulates NPR1 transcription, which in turn activates stress defense responses in plants under stress conditions (Belkhadir et al., 2004; Shi et al., 2015).

Likewise, SA-mediated regulation of other PTMs, i.e., ubiquitination of specific proteins, has been shown to stimulate plant growth and the immune system in response to biotic/abiotic stresses (Spoel et al., 2007; Miura et al., 2005). The most effective mechanism that promotes SA-mediated ubiquitination is the E3 ubiquitin ligase protein named NPR1 (Spoel et al., 2007; Fu et al., 2012). Increasing reports have delineated that NPR1 is a potent modulator of SA-mediated defense signaling and is responsible for its ubiquitination and degradation under normal growth conditions. This inhibition of NPR1 ubiquitination induces conformational stability of NPR1, thereby enhancing defense responses (Spoel et al., 2007; Fu et al., 2012).

SUMOylation is another type of PTM that modulates the activity of regulatory proteins in plants exposed to abiotic stresses (Lois and Lima, 2005; Miura et al., 2005). Several findings have contemplated the role of SA in modulating SUMOylation as part of signaling pathways, particularly in regulating the expression of specific proteins involved in plant defense responses (Lois and Lima, 2005; Kumar et al., 2012). For example, SUMOylation of NPR1 activates and mediates the translocation of NPR1 and TGA TFs in the nucleus, where it regulates the expression of defense-related genes (Kumar et al., 2012; Saleh et al., 2015).

NO-mediated post-translational modifications such as tyrosine nitration and S-nitrosylation also regulate plant growth and development under various stress conditions, including HT stress (Corpas et al., 2019; Corpas et al., 2019). Several findings have confirmed the role of NO signaling in stimulating the expression of defense-responsive proteins via a process known as tyrosine nitration by adding nitro group ($^{\cdot}NO_2$) to specific tyrosine residues in proteins (Corpas et al., 2021; Frungillo et al., 2014). Tyrosine nitration is mechanistically a covalent modification of proteins that instigates diverse effects on protein structure and functions. Increasing reports have contemplated that tyrosine nitration differentially regulates protein activity by modulating protein homeostasis, structural stability, biological activity, and subcellular localization that improves plant response under HT stress (Corpas et al., 2021; Corpas et al., 2019; Frungillo et al., 2014).

Likewise, S-nitrosation is also a NO-mediated PTM in plants which involves the modification of proteins by reversibly adding a nitric oxide radical ($^{\cdot}NO$) to a cysteine thiol group, thus generating SNO group (Corpas et al., 2021; Kolbert et al., 2021). Various studies have revealed that S-nitrosation is a redox-mediated PTM in plants that regulates protein function and activity under stress conditions (Lindermayr et al., 2010; Chen et al., 2010). In plants, S-nitrosation is believed to modulate the generation and scavenging of RNS, such as $ONOO^-$ and nitroxyl (HNO), enabling them to function as signaling molecules (Corpas et al., 2021; Frungillo et al., 2014). Upon acting as signaling molecules, these RNS regulate overall plant defense response by regulating the expression of defense-responsive proteins (Corpas et al., 2021; Frungillo et al., 2014). Furthermore, NO-mediated S-nitrosation also compellingly regulates several physiological processes, such as seed germination, root development, and stomatal regulation under HT stress (Corpas et al., 2021; Frungillo et al., 2014).

## Conclusion

SA and NO are binding signaling molecules that uniquely mediate various physiological and biochemical processes that enable plants to cope with high temperatures. Both SA and NO do it by triggering many responses, such as the activation of heat shock proteins and the induction of antioxidant defences. Furthermore, several studies have confirmed the interactive role of SA and NO and crosstalk with other phytohormones, suggesting the synergistic relationship in plant heat stress responses. SA can modulate NO biosynthesis, and NO, on the other hand, can stimulate SA production, thus highlighting the complex network of signaling events in plants. However, our understanding of the above notion between SA and NO in plant heat stress is still evolving. Nonetheless, these potential signaling molecules can be implemented in the breeding program, improving crop heat stress tolerance and productivity. Further research is needed to fully elucidate their roles and potential applications in agriculture and develop effective stress management strategies.

## References

AbdElgawad, H., Zinta, G., Hegab, M.M., Pandey, R., Asard, H. and Abuelsoud, W., 2016. High salinity induces different oxidative stress and antioxidant responses in maize seedlings organs. Frontiers in plant science, 7, p.276.

Arasimowicz, M. and Floryszak-Wieczorek, J., 2007. Nitric oxide as a bioactive signaling molecule in plant stress responses. Plant science, 172(5), pp.876-887.

Asgher, M., Per, T.S., Masood, A., Fatma, M., Freschi, L., Corpas, F.J. and Khan, N.A., 2017. Nitric oxide signaling and its crosstalk with other plant growth regulators in plant responses to abiotic stress. Environmental Science and Pollution Research, 24, pp.2273-2285.

Astier, J., Rasul, S., Koen, E., Manzoor, H., Besson-Bard, A., Lamotte, O., Jeandroz, S., Durner, J., Lindermayr, C. and Wendehenne, D., 2011. S-nitrosylation: an emerging post-translational protein modification in plants. Plant Science, 181(5), pp.527-533.

Bai, X., Yang, L., Tian, M., Chen, J., Shi, J., Yang, Y. and Hu, X., 2011. Nitric oxide enhances desiccation tolerance of recalcitrant Antiaris toxicaria seeds via protein S-nitrosylation and carbonylation. PLoS One, 6(6), p.e20714.

Beligni, M.V. and Lamattina, L., 2000. Nitric oxide stimulates seed germination and de-etiolation, and inhibits hypocotyl elongation, three light-inducible responses in plants. Planta, 210, pp.215-221.

Belkhadir, Y., Subramaniam, R. and Dangl, J.L., 2004. Plant disease resistance protein signaling: NBS–LRR proteins and their partners. Current opinion in plant biology, 7(4), pp.391-399.

Berr, A., Shafiq, S. and Shen, W.H., 2011. Histone modifications in transcriptional activation during plant development. Biochimica et Biophysica Acta (BBA)-Gene Regulatory Mechanisms, 1809(10), pp.567-576.

Besson-Bard, A., Courtois, C., Gauthier, A., Dahan, J., Dobrowolska, G., Jeandroz, S., Pugin, A. and Wendehenne, D., 2008. Nitric oxide in plants: production and cross-talk with Ca2+ signaling. Molecular Plant, 1(2), pp.218-228.

Bethke, P.C., Libourel, I.G., Aoyama, N., Chung, Y.Y., Still, D.W. and Jones, R.L., 2007. The Arabidopsis aleurone layer responds to nitric oxide, gibberellin, and abscisic acid and is sufficient and necessary for seed dormancy. Plant physiology, 143(3), pp.1173-1188.

Bhattacharjee, S. and Bhattacharjee, S., 2019. Exploring Oxidative Stress in Plants: Proteomic and Genomic Approaches. Reactive Oxygen Species in Plant Biology, pp.155-187.

Birkenbihl, R.P., Liu, S. and Somssich, I.E., 2017. Transcriptional events defining plant immune responses. Current Opinion in Plant Biology, 38, pp.1-9.

Cao, H., Glazebrook, J., Clarke, J.D., Volko, S. and Dong, X., 1997. The Arabidopsis NPR1 gene that controls systemic acquired resistance encodes a novel protein containing ankyrin repeats. Cell, 88(1), pp.57-63.

Chen, F., Wang, F., Sun, H., Cai, Y., Mao, W., Zhang, G., Vincze, E. and Wu, F., 2010. Genotype-dependent effect of exogenous nitric oxide on Cd-induced changes in antioxidative metabolism, ultrastructure, and photosynthetic performance in barley seedlings (Hordeum vulgare). Journal of Plant Growth Regulation, 29, pp.394-408.

Chen, J., Xiao, Q., Wu, F., Dong, X., He, J., Pei, Z. and Zheng, H., 2010. Nitric oxide enhances salt secretion and Na+ sequestration in a mangrove plant, Avicennia marina, through increasing the expression of H+-ATPase and Na+/H+ antiporter under high salinity. Tree physiology, 30(12), pp.1570-1585.

Christou, A., Filippou, P., Manganaris, G.A. and Fotopoulos, V., 2014. Sodium hydrosulfide induces systemic thermotolerance to strawberry plants through transcriptional regulation of heat shock proteins and aquaporin. BMC Plant Biology, 14(1), pp.1-11.

Corpas, F.J., Del Río, L.A. and Palma, J.M., 2019. Impact of nitric oxide (NO) on the ROS metabolism of peroxisomes. Plants, 8(2), p.37.

Corpas, F.J., González-Gordo, S. and Palma, J.M., 2021. Protein nitration: A connecting bridge between nitric oxide (NO) and plant stress. Plant Stress, 2, p.100026.

Del Castello, F., Nejamkin, A., Cassia, R., Correa-Aragunde, N., Fernández, B., Foresi, N., Lombardo, C., Ramirez, L. and Lamattina, L., 2019. The era of nitric oxide in plant biology: Twenty years tying up loose ends. Nitric Oxide, 85, pp.17-27.

Devireddy, A.R., Tschaplinski, T.J., Tuskan, G.A., Muchero, W. and Chen, J.G., 2021. Role of reactive oxygen species and hormones in plant responses to temperature changes. International Journal of Molecular Sciences, 22(16), p.8843.

Dixon, R.A., Achnine, L., Kota, P., Liu, C.J., Reddy, M.S. and Wang, L., 2002. The phenylpropanoid pathway and plant defence—a genomics perspective. Molecular plant pathology, 3(5), pp.371-390.

Domingos, P., Prado, A.M., Wong, A., Gehring, C. and Feijo, J.A., 2015. Nitric oxide: a multitasked signaling gas in plants. Molecular plant, 8(4), pp.506-520.

Dong, X., Yang, Y., Zhang, Z., Xiao, Z., Bai, X., Gao, J., Hur, Y., Hao, S. and He, F., 2019. Genome-wide identification of WRKY genes and their response to cold stress in *Coffea canephora*. Forests, 10(4), p.335.

Dong, Z., Yu, Y., Li, S., Wang, J., Tang, S. and Huang, R., 2016. Abscisic acid antagonizes ethylene production through the ABI4-mediated transcriptional repression of ACS4 and ACS8 in Arabidopsis. Molecular plant, 9(1), pp.126-135.

Durrant, W.E. and Dong, X., 2004. Systemic acquired resistance. Annu. Rev. Phytopathol., 42, pp.185-209.

Fan, H.F., Du, C.X. and Guo, S.R., 2013. Nitric oxide enhances salt tolerance in cucumber seedlings by regulating free polyamine content. Environmental and Experimental Botany, 86, pp.52-59.

Feechan, A., Kwon, E., Yun, B.W., Wang, Y., Pallas, J.A. and Loake, G.J., 2005. A central role for S-nitrosothiols in plant disease resistance. Proceedings of the National Academy of Sciences, 102(22), pp.8054-8059.

Feki, K., Tounsi, S., Mrabet, M., Mhadhbi, H. and Brini, F., 2021. Recent advances in physiological and molecular mechanisms of heavy metal accumulation in plants. Environmental Science and Pollution Research, pp.1-20.

Fonseca, A., Urzúa, T., Jelenska, J., Sbarbaro, C., Seguel, A., Duarte, Y., Greenberg, J.T., Holuigue, L., Blanco-Herrera, F. and Herrera-Vásquez, A., 2022. The TGA Transcription Factors from Clade II Negatively Regulate the Salicylic Acid Accumulation in Arabidopsis. International Journal of Molecular Sciences, 23(19), p.11631.

Fraser, C.M., Chapple, C. 2011. The phenylpropanoid pathway in Arabidopsis. Arabidopsis Book, (9), p.e0152.

Freschi, L., 2013. Nitric oxide and phytohormone interactions: current status and perspectives. Frontiers in plant science, 4, p.398.

Frungillo, L., Skelly, M.J., Loake, G.J., Spoel, S.H. and Salgado, I., 2014. S-nitrosothiols regulate nitric oxide production and storage in plants through the nitrogen assimilation pathway. Nature Communications, 5(1), p.5401.

Fu, Z.Q., Yan, S., Saleh, A., Wang, W., Ruble, J., Oka, N., Mohan, R., Spoel, S.H., Tada, Y., Zheng, N. and Dong, X., 2012. NPR3 and NPR4 are receptors for the immune signal salicylic acid in plants. Nature, 486(7402), pp.228-232.

Fujita, M., Fujita, Y., Maruyama, K., Seki, M., Hiratsu, K., Ohme-Takagi, M., Tran, L.S.P., Yamaguchi-Shinozaki, K. and Shinozaki, K., 2004. A dehydration-induced NAC protein, RD26, is involved in a novel ABA-dependent stress-signaling pathway. The Plant Journal, 39(6), pp.863-876.

Gao, F., Zhou, J., Deng, R.Y., Zhao, H.X., Li, C.L., Chen, H., Suzuki, T., Park, S.U. and Wu, Q., 2017. Overexpression of a tartary buckwheat R2R3-MYB transcription factor gene, FtMYB9, enhances tolerance to drought and salt stresses in transgenic Arabidopsis. Journal of Plant Physiology, 214, pp.81-90.

García-Mata, C. and Lamattina, L., 2001. Nitric oxide induces stomatal closure and enhances the adaptive plant responses against drought stress. Plant Physiology, 126(3), pp.1196-1204.

Garcion, C., Lohmann, A., Lamodière, E., Catinot, J., Buchala, A., Doermann, P. and Métraux, J.P., 2008. Characterization and biological function of the ISOCHORISMATE SYNTHASE2 gene of Arabidopsis. Plant physiology, 147(3), pp.1279-1287.

Gautam, H., Fatma, M., Sehar, Z., Mir, I.R. and Khan, N.A., 2022. Hydrogen sulfide, ethylene, and nitric oxide regulate redox homeostasis and protect photosynthetic metabolism under high temperature stress in rice plants. Antioxidants, 11(8), p.1478.

Gill, S.S., Anjum, N.A., Hasanuzzaman, M., Gill, R., Trivedi, D.K., Ahmad, I., Pereira, E. and Tuteja, N., 2013. Glutathione and glutathione reductase: a boon in disguise for plant abiotic stress defense operations. Plant Physiology and Biochemistry, 70, pp.204-212.

Guo, F.Q., Okamoto, M. and Crawford, N.M., 2003. Identification of a plant nitric oxide synthase gene involved in hormonal signaling. Science, 302(5642), pp.100-103.

Gupta, K.J. and Igamberdiev, A.U., 2016. Reactive nitrogen species in mitochondria and their implications in plant energy status and hypoxic stress tolerance. Frontiers in Plant Science, 7, p.369.

Horváth, E., Szalai, G. and Janda, T., 2007. Induction of abiotic stress tolerance by salicylic acid signaling. Journal of Plant Growth Regulation, 26, pp.290-300.

Hu, Y., Jiang, Y., Han, X., Wang, H., Pan, J. and Yu, D., 2017. Jasmonate regulates leaf senescence and tolerance to cold stress: crosstalk with other phytohormones. Journal of Experimental Botany, 68(6), pp.1361-1369.

Isner, J.C. and Maathuis, F.J., 2016. cGMP signaling in plants: from enigma to main stream. Functional Plant Biology, 45(2), pp.93-101.

Johnson, C., Boden, E. and Arias, J., 2003. Salicylic acid and NPR1 induce the recruitment of trans-activating TGA factors to a defense gene promoter in Arabidopsis. The Plant Cell, 15(8), pp.1846-1858.

Katiyar, A., Smita, S., Lenka, S.K., Rajwanshi, R., Chinnusamy, V. and Bansal, K.C., 2012. Genome-wide classification and expression analysis of MYB transcription factor families in rice and Arabidopsis. BMC genomics, 13, pp.1-19.

Kazan, K. and Manners, J.M., 2013. MYC2: the master in action. Molecular plant, 6(3), pp.686-703.

Khan, M., Al Azzawi, T.N.I., Ali, S., Yun, B.W. and Mun, B.G., 2023. Nitric Oxide, a Key Modulator in the Alleviation of Environmental Stress-Mediated Damage in Crop Plants: A Meta-Analysis. Plants, 12(11), p.2121..

Khan, M.I.R., Fatma, M., Per, T.S., Anjum, N.A. and Khan, N.A., 2015. Salicylic acid-induced abiotic stress tolerance and underlying mechanisms in plants. Frontiers in plant science, 6, p.462.

Khraiwesh, B., Zhu, J.K. and Zhu, J., 2012. Role of miRNAs and siRNAs in biotic and abiotic stress responses of plants. Biochimica et Biophysica Acta (BBA)-Gene Regulatory Mechanisms, 1819(2), pp.137-148.

Khurana, N., Sharma, N. and Khurana, P., 2017. Overexpression of a heat stress inducible, wheat myo-inositol-1-phosphate synthase 2 (TaMIPS2) confers tolerance to various abiotic stresses in *Arabidopsis thaliana*. Agri Gene, 6, pp.24-30.

Kolbert, Z., Lindermayr, C. and Loake, G.J., 2021. The role of nitric oxide in plant biology: current insights and future perspectives. Journal of Experimental Botany, 72(3), pp.777-780.

Kumar, R., Agarwal, P., Tyagi, A.K. and Sharma, A.K., 2012. Genome-wide investigation and expression analysis suggest diverse roles of auxin-responsive GH3 genes during development and response to different stimuli in tomato (*Solanum lycopersicum*). Molecular Genetics and Genomics, 287, pp.221-235.

Lefevere, H., Bauters, L. and Gheysen, G., 2020. Salicylic acid biosynthesis in plants. Frontiers in plant science, 11, p.338.

Li, Y., Shan, X., Jiang, Z., Zhao, L. and Jin, F., 2021. Genome-wide identification and expression analysis of the GA2ox gene family in maize (Zea mays L.) under various abiotic stress conditions. Plant Physiology and Biochemistry, 166, pp.621-633.

Lindermayr, C., Sell, S., Müller, B., Leister, D. and Durner, J., 2010. Redox regulation of the NPR1-TGA1 system of Arabidopsis thaliana by nitric oxide. The Plant Cell, 22(8), pp.2894-2907.

Liu, J., He, Y., Amasino, R. and Chen, X., 2004. siRNAs targeting an intronic transposon in the regulation of natural flowering behavior in Arabidopsis. Genes & development, 18(23), pp.2873-2878.

Liu, J., Qiu, G., Liu, C., Li, H., Chen, X., Fu, Q., Lin, Y. and Guo, B., 2022. Salicylic acid, a multifaceted hormone, combats abiotic stresses in plants. Life, 12(6), p.886.

Liu, L., Hausladen, A., Zeng, M., Que, L., Heitman, J. and Stamler, J.S., 2001. A metabolic enzyme for S-nitrosothiol conserved from bacteria to humans. Nature, 410(6827), pp.490-494.

Liu, L., Sonbol, F.M., Huot, B., Gu, Y., Withers, J., Mwimba, M., Yao, J., He, S.Y. and Dong, X., 2016. Salicylic acid receptors activate jasmonic acid signaling through a non-canonical pathway to promote effector-triggered immunity. Nature communications, 7(1), p.13099.

Lois, L.M. and Lima, C.D., 2005. Structures of the SUMO E1 provide mechanistic insights into SUMO activation and E2 recruitment to E1. The EMBO journal, 24(3), pp.439-451.

Lorenzo, O., Chico, J.M., Saénchez-Serrano, J.J. and Solano, R., 2004. JASMONATE-INSENSITIVE1 encodes a MYC transcription factor essential to discriminate between different jasmonate-regulated defense responses in Arabidopsis. The Plant Cell, 16(7), pp.1938-1950.

Lozano-Juste, J. and León, J., 2011. Nitric oxide regulates DELLA content and PIF expression to promote photomorphogenesis in Arabidopsis. Plant Physiology, 156(3), pp.1410-1423.

Luna, E., Bruce, T.J., Roberts, M.R., Flors, V. and Ton, J., 2012. Next-generation systemic acquired resistance. Plant physiology, 158(2), pp.844-853.

Luo, M., Luo, H., Li, H., Liu, D. and Lu, H., 2020. Single-stranded DNA-binding proteins in plant telomeres. International Journal of Biological Macromolecules, 165, pp.1463-1467.

Luo, M., Tai, R., Yu, C.W., Yang, S., Chen, C.Y., Lin, W.D., Schmidt, W. and Wu, K., 2015. Regulation of flowering time by the histone deacetylase HDA 5 in A rabidopsis. The Plant Journal, 82(6), pp.925-936.

Maeda, H. and Dudareva, N., 2012. The shikimate pathway and aromatic amino acid biosynthesis in plants. Annual review of plant biology, 63, pp.73-105.

Manjunatha, G., Lokesh, V. and Neelwarne, B., 2010. Nitric oxide in fruit ripening: trends and opportunities. Biotechnology advances, 28(4), pp.489-499.

Mao, P., Jin, X., Bao, Q., Mei, C., Zhou, Q., Min, X. and Liu, Z., 2020. WRKY transcription factors in Medicago sativa L.: Genome-wide identification and expression analysis under abiotic stress. DNA and Cell Biology, 39(12), pp.2212-2225.

Mauch-Mani, B. and Slusarenko, A.J., 1996. Production of salicylic acid precursors is a major function of phenylalanine ammonia-lyase in the resistance of Arabidopsis to Peronospora parasitica. The Plant Cell, 8(2), pp.203-212.

Meng, Y., Shao, C., Wang, H. and Chen, M., 2011. The regulatory activities of plant microRNAs: a more dynamic perspective. Plant Physiology, 157(4), pp.1583-1595.

Mhamdi, A., Noctor, G. and Baker, A., 2012. Plant catalases: peroxisomal redox guardians. Archives of Biochemistry and Biophysics, 525(2), pp.181-194.

Mishina, T.E. and Zeier, J., 2007. Pathogen-associated molecular pattern recognition rather than development of tissue necrosis contributes to bacterial induction of systemic acquired resistance in Arabidopsis. The Plant Journal, 50(3), pp.500-513.

Miura, K. and Tada, Y., 2014. Regulation of water, salinity, and cold stress responses by salicylic acid. Frontiers in plant science, 5, p.4.

Miura, K., Rus, A., Sharkhuu, A., Yokoi, S., Karthikeyan, A.S., Raghothama, K.G., Baek, D., Koo, Y.D., Jin, J.B., Bressan, R.A. and Yun, D.J., 2005. The Arabidopsis SUMO E3 ligase SIZ1 controls phosphate deficiency responses. Proceedings of the National Academy of Sciences, 102(21), pp.7760-7765.

Molinier, J., Ries, G., Zipfel, C. and Hohn, B., 2006. Transgeneration memory of stress in plants. Nature, 442(7106), pp.1046-1049.

Nabi, R.B.S., Tayade, R., Hussain, A., Kulkarni, K.P., Imran, Q.M., Mun, B.G. and Yun, B.W., 2019. Nitric oxide regulates plant responses to drought, salinity, and heavy metal stress. Environmental and Experimental Botany, 161, pp.120-133.

Neill, S.J., Desikan, R. and Hancock, J.T., 2003. Nitric oxide signaling in plants. New Phytologist, 159(1), pp.11-35.

Neill, S.J., Desikan, R., Clarke, A. and Hancock, J.T., 2002. Nitric oxide is a novel component of abscisic acid signaling in stomatal guard cells. Plant physiology, 128(1), pp.13-16. Baudouin, E. and Hancock, J.T., 2014. Nitric oxide signaling in plants. Frontiers in Plant Science, 4, p.553.

Parankusam, S., Adimulam, S.S., Bhatnagar-Mathur, P. and Sharma, K.K., 2017. Nitric oxide (NO) in plant heat stress tolerance: current knowledge and perspectives. Frontiers in Plant Science, 8, p.1582.

Pauwels, L., Barbero, G.F., Geerinck, J., Tilleman, S., Grunewald, W., Pérez, A.C., Chico, J.M., Bossche, R.V., Sewell, J., Gil, E. and Garcia-Casado, G., 2010. NINJA connects the co-repressor TOPLESS to jasmonate signaling. Nature, 464(7289), pp.788-791.

Prakash, V., Singh, V.P., Tripathi, D.K., Sharma, S. and Corpas, F.J., 2021. Nitric oxide (NO) and salicylic acid (SA): A framework for their relationship in plant development under abiotic stress. Plant Biology, 23, pp.39-49.

Rai, K.K., Pandey, N. and Rai, S.P., 2020. Salicylic acid and nitric oxide signaling in plant heat stress. Physiologia plantarum, 168(2), pp.241-255.

Rai, K.K., Pandey, N., Rai, N., Rai, S.K. and Pandey-Rai, S., 2021. Salicylic acid and nitric oxide: insight into the transcriptional regulation of their metabolism and regulatory functions in plants. Frontiers in Agronomy, 3, p.781027.

Rivas-San Vicente, M. and Plasencia, J., 2011. Salicylic acid beyond defence: its role in plant growth and development. Journal of experimental botany, 62(10), pp.3321-3338.

Rockel, P., Strube, F., Rockel, A., Wildt, J. and Kaiser, W.M., 2002. Regulation of nitric oxide (NO) production by plant nitrate reductase in vivo and in vitro. Journal of experimental botany, 53(366), pp.103-110.

Ruan, Y.L., Llewellyn, D.J. and Furbank, R.T., 2001. The control of single-celled cotton fiber elongation by developmentally reversible gating of plasmodesmata and coordinated expression of sucrose and K+ transporters and expansin. The Plant Cell, 13(1), pp.47-60.

Rubio-Somoza, I. and Weigel, D., 2011. MicroRNA networks and developmental plasticity in plants. Trends in plant science, 16(5), pp.258-264.

Saleh, A., Withers, J., Mohan, R., Marqués, J., Gu, Y., Yan, S., Zavaliev, R., Nomoto, M., Tada, Y. and Dong, X., 2015. Posttranslational modifications of the master transcriptional regulator NPR1 enable dynamic but tight control of plant immune responses. Cell host & microbe, 18(2), pp.169-182.

Savvides, A., Ali, S., Tester, M. and Fotopoulos, V., 2016. Chemical priming of plants against multiple abiotic stresses: mission possible?. Trends in plant science, 21(4), pp.329-340.

Shivaraj, S.M., Vats, S., Bhat, J.A., Dhakte, P., Goyal, V., Khatri, P., Kumawat, S., Singh, A., Prasad, M., Sonah, H. and Sharma, T.R., 2020. Nitric oxide and hydrogen sulfide crosstalk during heavy metal stress in plants. Physiologia Plantarum, 168(2), pp.437-455.

Spoel, S.H., Johnson, J.S. and Dong, X., 2007. Regulation of tradeoffs between plant defenses against pathogens with different lifestyles. Proceedings of the National Academy of Sciences, 104(47), pp.18842-18847.

Spoel, S.H., Koornneef, A., Claessens, S.M., Korzelius, J.P., Van Pelt, J.A., Mueller, M.J., Buchala, A.J., Métraux, J.P., Brown, R., Kazan, K. and Van Loon, L.C., 2003. NPR1 modulates cross-talk between salicylate-and jasmonate-dependent defense pathways through a novel function in the cytosol. The Plant Cell, 15(3), pp.760-770.

Terrile, M.C., París, R., Calderón-Villalobos, L.I., Iglesias, M.J., Lamattina, L., Estelle, M. and Casalongué, C.A., 2012. Nitric oxide influences auxin signaling through S-nitrosylation of the Arabidopsis TRANSPORT INHIBITOR RESPONSE 1 auxin receptor. The Plant Journal, 70(3), pp.492-500.

Ushimaru, T., Nakagawa, T., Fujioka, Y., Daicho, K., Naito, M., Yamauchi, Y., Nonaka, H., Amako, K., Yamawaki, K. and Murata, N., 2006. Transgenic Arabidopsis plants expressing the rice dehydroascorbate reductase gene are resistant to salt stress. Journal of plant physiology, 163(11), pp.1179-1184.

Van Verk, M.C., Gatz, C. and Linthorst, H.J., 2009. Transcriptional regulation of plant defense responses. Advances in Botanical Research, 51, pp.397-438.

Vlot, A.C., Dempsey, D.M.A. and Klessig, D.F., 2009. Salicylic acid, a multifaceted hormone to combat disease. Annual review of phytopathology, 47, pp.177-206.

Wang, K.L.C., Li, H. and Ecker, J.R., 2002. Ethylene biosynthesis and signaling networks. The plant cell, 14(suppl_1), pp.S131-S151.
Widhalm, J.R., Jaini, R., Morgan, J.A. and Dudareva, N., 2015. Rethinking how volatiles are released from plant cells. Trends in plant science, 20(9), pp.545-550.
Wu, X., Zhu, W., Zhang, H., Ding, H. and Zhang, H.J., 2011. Exogenous nitric oxide protects against salt-induced oxidative stress in the leaves from two genotypes of tomato (*Lycopersicom esculentum* Mill.). Acta physiologiae plantarum, 33, pp.1199-1209.
Wu, Y., Zhang, D., Chu, J.Y., Boyle, P., Wang, Y., Brindle, I.D., De Luca, V. and Després, C., 2012. The Arabidopsis NPR1 protein is a receptor for the plant defense hormone salicylic acid. Cell reports, 1(6), pp.639-647.
Xu, X.T., Jin, X., Liao, W.B., Dawuda, M.M., Li, X.P., Wang, M., Niu, L.J., Ren, P.J. and Zhu, Y.C., 2017. Nitric oxide is involved in ethylene-induced adventitious root development in cucumber (Cucumis sativus L.) explants. Scientia Horticulturae, 215, pp.65-71.
Xuan, Y., Zhou, S., Wang, L., Cheng, Y. and Zhao, L., 2010. Nitric oxide functions as a signal and acts upstream of AtCaM3 in thermotolerance in Arabidopsis seedlings. Plant Physiology, 153(4), pp.1895-1906.
Yu, C.W., Liu, X., Luo, M., Chen, C., Lin, X., Tian, G., Lu, Q., Cui, Y. and Wu, K., 2011. HISTONE DEACETYLASE6 interacts with FLOWERING LOCUS D and regulates flowering in Arabidopsis. Plant Physiology, 156(1), pp.173-184.
Yu, M., Lamattina, L., Spoel, S.H. and Loake, G.J., 2014. Nitric oxide function in plant biology: a redox cue in deconvolution. New phytologist, 202(4), pp.1142-1156.
Yun, B.W., Feechan, A., Yin, M., Saidi, N.B., Le Bihan, T., Yu, M., Moore, J.W., Kang, J.G., Kwon, E., Spoel, S.H. and Pallas, J.A., 2011. S-nitrosylation of NADPH oxidase regulates cell death in plant immunity. Nature, 478(7368), pp.264-268.
Zhang, L., Wang, K., Han, Y., Yan, L., Zheng, Y., Bi, Z., Zhang, X., Zhang, X. and Min, D., 2022. Genome-wide analysis of the VQ motif-containing gene family and expression profiles during phytohormones and abiotic stresses in wheat (*Triticum aestivum* L.). BMC genomics, 23(1), pp.1-20.
Zhang, X., Yazaki, J., Sundaresan, A., Cokus, S., Chan, S.W.L., Chen, H., Henderson, I.R., Shinn, P., Pellegrini, M., Jacobsen, S.E. and Ecker, J.R., 2006. Genome-wide high-resolution mapping and functional analysis of DNA methylation in Arabidopsis. Cell, 126(6), pp.1189-1201.

Zhang, Y., Wang, L., Liu, Y., Zhang, Q., Wei, Q. and Zhang, W., 2006. Nitric oxide enhances salt tolerance in maize seedlings through increasing activities of proton-pump and Na+/H+ antiport in the tonoplast. Planta, 224, pp.545-555.

Zhou, B., Guo, Z., Xing, J. and Huang, B., 2005. Nitric oxide is involved in abscisic acid-induced antioxidant activities in Stylosanthes guianensis. Journal of Experimental Botany, 56(422), pp.3223-3228.

# CHAPTER 5

# SALICYLIC ACID AND NITRIC OXIDE: SIGNALING IN HEAT STRESS


## Abstract

During heat stress exposure, plants employ various signaling cascades to provoke stress defence response via activating downstream signaling genes. These signaling cascades also trigger stress-adaptive reactions by modulating the interplay between salicylic acid (SA) and nitric oxide (NO) signaling, thus acting as a critical regulator of thermotolerance in plant heat stress responses. SA regulates HT stress response in plants by acting as transcriptional activation of various transcription factors (TFs), secondary metabolites, and stress-responsive genes/proteins. NO signaling under HT stress functions as a gaseous signaling molecule that regulates stomatal closure, thus preventing excessive water loss through transpiration. Furthermore, SA can induce NO production via a positive feedback loop, and their interactive role leads to the activation of heat shock proteins (HSPs). Therefore, this chapter provides the latest information regarding the SA and NO signaling in plant heat stress. Further, the chapter also uncovers the novel signaling components that are responsible for SA/NO perception and orchestration of HT stress response in plants. Lastly, the mechanism underlying the interaction among SA, NO and other signaling molecules is briefly discussed.



**Keywords**: Signaling, Perception, Receptor protein, Interaction, Stress


## Introduction

The phytohormone SA and gaseous molecule NO act as signaling molecules in plants, involved in the regulation of various physiological and defense responses (Vlot et al., 2009; Arasimowicz-Jelonek and Floryszak-Wieczorek, 2014). Specific receptors or sensors perceive the SA signaling in plants specifically interact and bind to SA to initiate the signaling cascades (Mishina and Zeier, 2007; Tillmann, 2021). Scientists working in the plant scientific community have a biased opinion about the exact mechanism of SA perception (Ding et al., 2018). Nonetheless, they have identified several proteins which potentially function as SA receptors or their regulatory components are involved in SA perception (Xu et al., 2022; Chen et al., 2021). SA functions to regulate plant defense against biotic and abiotic stresses by activating systemic acquired resistance (SAR). Furthermore, SA-mediated activation of SAR activates the transcription of several defense-related genes, exaggerating the synthesis of secondary metabolites, thereby strengthening cell walls and plant innate immune responses (Dempsey et al., 2011; Fu and Dong, 2013; Arasimowicz and Floryszak-Wieczorek, 2007).

The Nonexpresser of Pathogenesis-Related Genes 1 (NPR1) is believed to be the central regulator of SA signaling and also plays a profound role in SA perception (Fu et al., 2020). When the SA level is below the threshold, i.e. under normal conditions, the NPR1 remains an oligomer in the cytoplasm. However, when the plant experiences stress anomaly, the upsurge in SA synthesis is perceived by NPR1, which is then converted into a monomer and translocated into the nucleus to stimulate defense response (Chamizo-Ampudia et al., 2017). Conversely, NPR3 and NPR4 are analogous to NPR1 but suppress SA perception and signaling. The NPR3 and NPR4 interact and bind SA receptors, which then degrade NPR1, thus negatively regulating SA signaling (Song et al., 2008). Likewise, Enhanced Disease Susceptibility 1 (EDS1) and Systemic Acquired Resistance-Deficient 1 (SARD1) are other components that regulate SA perception and signaling (Cui et al., 2015; Xu et al., 2022).

The mechanism of perception of NO in plants is still not fully understood and is an area of active research (Astier and Lindermayr, 2012; Gupta et al., 2021). Nonetheless, researchers have identified some proteins which may mechanistically involve in the perception and signaling of NO in plants (Gupta et al., 2011). The enzyme guanylate cyclases that catalyze the conversion of GTP to cGMP also readily function as NO receptors in plants. The enzyme perceives and activates NO signaling in plants by acting as a downstream signaling molecule and stimulating cGMP production (Gibbs et al., 2014; del Rıo et al., 2004). Similarly, the enzyme S-nitroso glutathione reductase (GSNOR) has also been reported to function as a NO sensor which activates NO signaling upon interacting (Keszler et al., 2010; Song et al., 2022). Furthermore, GSNOR also performs catalytic degradation of GSNO to produce NO. Thioredoxins (Trx) are a class of redox-active proteins that are specifically involved in the maintenance of redox homeostasis in plants under stress conditions

(Keech et al., 2017). Studies have also reported that it also functions as a NO sensor, and its interaction with NO induces a conformational change in Trx and transmits a signal to downstream targets. Since understanding NO perception and signaling in plants is still evolving, unraveling its crosstalk with other signaling pathways may provide valuable insight into the intricate mechanism of NO perception in plants.

## 1. The EDS1 PAD4 SAG101 triad

Enhanced Disease Susceptibility 1 (EDS1) is profoundly involved in regulating many growth and defense signaling pathways in plants by modulating SA. It also plays a significant role in regulating the activation of SA signaling, thereby conferring biotic and abiotic stress tolerance in plants (Wiermer et al., 2005; Su et al., 2014). In recent years, tremendous work has been done to unravel the potential mechanism behind SA signaling. In this continuation, several findings have reported that SA signaling in plants is explicitly mediated by EDS1, which is a multifunctional protein involved in the activation of SA-dependent defense responses (Feys et al., 2001; Cui et al., 2017). Further, studies have also confirmed that EDS1 regulates SA perception and signaling after interacting and forming a complex with other proteins, such as PAD4 (Phytoalexin Deficient 4) and SAG101 (Senescence-Associated Gene 101) (Aarts et al., 1998; Waterhouse et al., 1998). The complex formed, i.e., EDS1-PAD4 and EDS1-SAG101, serve as an imperative junction for the mechanical regulation of SA-mediated defense response in plants (Feys et al., 2001; Waterhouse et al., 1998).

In addition, the EDS1 protein is also known to regulate different aspects of plant growth, development, and defense response (Wiermer et al., 2005; Sue et al. 2014). For instance, SA signaling interacts and activates the jasmonic acid (JA) pathway and the ethylene (ET) pathway (Pieterse et al., 2012; Kazan and Manners, 2013). The crosstalk between SA-JA and ethylene allows plants to rewire their defense strategies by activating different defense signaling cascades and to create systemic memory based on the biotic/abiotic stress they encounter (Mauch-Mani and Mauch, 2005; Pieterse et al., 2012). Correspondingly, the PAD4-mediated activation of SA signaling in plants involves the regulation of various downstream signaling genes that are an integral component of the gene cluster of stress defense response (Figure 1) (Wiermer et al., 2005). However, researchers have not identified a specific gene group related to PAD4 signaling. They hypothesized that PAD4 activates SA signaling and perception upon interacting with other genes and transcription factors to regulate plant tolerance to heat stress. Research findings suggest that PAD4 interacts and forms a complex with EDS1, which acts as a molecular junction between SA and JA defense pathways. Additionally, PAD4 has also been implicated in stimulating the production of SA in plants by activating the transcription of SA biosynthetic genes (Chen et al., 2021). Intriguingly, studies have also corroborated the involvement of PAD4 in the development and execution of hypersensitive response (HR), a type of cell apoptosis which is triggered during pathogen attack (Lorrain et al., 2003).

SAG101 protein has also been shown to regulate SA signaling and perception in plants exposed to HT stress (Tillmann, 2021). Further, recent findings have attested to its role in activating SA-mediated defense responses leading to modulation of plant innate immune response (Coqueiro et al., 2015). Studies have suggested that SAG101 also forms a complex with EDS, thereby acting as a central regulator of SA signaling in plants via interacting with JA and ethylene (Feys et al., 2001; Wiermer et al., 2005). SAG101 exerts regulatory functions by activating its nucleocytoplasmic shuttling (Cui et al., 2017). It is generally believed that SAG101 is predominantly present in the cytoplasm; however, upon SA stimulation, SAG101 migrates to the nucleus to unravel its full regulatory functions (Wiermer et al., 2005).

## 2. Molecular signaling of SA

The molecular signaling of salicylic acid in plants involves intricate signaling of several downstream signaling genes, which strengthen plant defense response under extreme environments (Chen et al., 2021). The SA signaling is incepted by the perception of SA through specific receptor proteins or sensors involving NPR1 and TGA transcription factors (Table 1) (Wu et al., 2012; Wiermer et al., 2005). Once activated, these NPR1 and TGA TFs stimulate SA biosynthesis via a positive feedback loop, which upregulates ISOCHORISMATE SYNTHASE 1 (ICS1) and PHYTOALEXIN DEFICIENT 4 (PAD4) genes of the SA biosynthetic pathway (Aygören et al., 2023). This section will uncover the recent advancement in unraveling the molecular basis of SA signaling in plants.

**Table 1** Biochemical and molecular signaling of salicylic acid (SA) and nitric oxide (NO) in plants under heat stress

| Signaling of SA and NO in Plants under Heat Stress | Mechanism involved | Genes/Proteins Involved in SA Signaling | Genes/Proteins Involved in NO Signaling | References |
|---|---|---|---|---|
| **Induction of SA and NO production** | Heat stress triggers the synthesis and accumulation of salicylic acid (SA) and nitric oxide (NO) in plants. SA and NO act as important signaling molecules, mediating various physiological and molecular responses to heat stress. | NPR1 (Nonexpressor of Pathogenesis-Related genes 1), PAL (Phenylalanine Ammonia-Lyase), EDS1 (Enhanced Disease Susceptibility 1), PAD4 (Phytoalexin Deficient 4) | NIA1/2 (Nitrate Reductase 1/2), NR (Nitrate Reductase), NRX1 (Nitrate Reductase Regulator 1), NRX2 (Nitrate Reductase Regulator 2), NOA1 (Nitric Oxide Associated 1), NRAMP3 (Natural Resistance-Associated Macrophage Protein 3) | Wang et al. (2020) |
| **Activation of Defense Responses** | SA signaling pathway activates defense responses against heat stress in plants. SA induces the expression of heat shock proteins (HSPs), antioxidant enzymes, and other stress-responsive genes. | PR (Pathogenesis-Related) genes (e.g., PR1, PR2, PR5), HSPs (Heat Shock Proteins), APX (Ascorbate Peroxidase), CAT (Catalase), SOD (Superoxide Dismutase), GST (Glutathione S-Transferase), WRKY transcription factors | NIA1/2 (Nitrate Reductase 1/2), NR (Nitrate Reductase), NRX1 (Nitrate Reductase Regulator 1), NRX2 (Nitrate Reductase Regulator 2), NOS (Nitric Oxide Synthase), ROS (Reactive Oxygen Species), MAPKs (Mitogen-Activated Protein Kinases), APX (Ascorbate Peroxidase), CAT (Catalase), SOD (Superoxide Dismutase), GST (Glutathione S-Transferase), WRKY transcription factors | Neill et al. (2008) |

| | | | | |
|---|---|---|---|---|
| **Crosstalk with Other Signaling Pathways** | SA and NO signaling pathways interact with other signaling pathways to integrate and coordinate responses to heat stress in plants. | JA (Jasmonic Acid) signaling pathway: LOX (Lipoxygenase), AOS (Allene Oxide Synthase), AOC (Allene Oxide Cyclase), COI1 (Coronatine Insensitive 1) | ROS (Reactive Oxygen Species) signaling pathway: RBOHD/F (Respiratory Burst Oxidase Homolog D/F), NADPH oxidase, MAPKs (Mitogen-Activated Protein Kinases) | Van Breusegem and Dat (2006) |

## 2.1. NPR1-dependent signaling

Increasing literature has affirmed that NPR1 is a critical component of SA signaling that regulates various growth and stress defense responses in plants (Mou et al., 2003). NPR1 modulates SA biosynthesis and signaling by acting as a transcriptional co-activator of several downstream signaling genes (Wang et al., 2006; Germain et al., 2010). For instance, studies have shown that NIM1-interacting (NIMIN) genes, such as NIMIN1 and NIMIN2, encode proteins that specifically interact with NPR1, thus regulating its activity by modulating its oligomeric conformation (Fu et al., 2021; Song et al., 2008). NIMIN1 positively regulates NPR1-mediated SA signaling, while NIMIN2 functions antagonistically by inhibiting NPR1-mediated SA signaling (Li et al., 2004; Wu et al., 2012). Similarly, TGA TFs also play a substantial role in regulating NPR1-mediated SA signaling in plants. TGA TFs exert this regulatory function by interacting with GCC-box cis-regulatory elements present in the promoter regions of various defense-related genes (Wu et al., 2012).

WRKY TFs are also one such protein that interacts and binds NPR1 proteins to stimulate SA biosynthesis and perception in plants exposed to biotic/abiotic stresses (Li et al., 2004). WRKY proteins stimulate NPR1-mediated SA signaling after binding to the W-box element, which is a cis-regulatory element found in the promoter of several genes (Pandey and Somssich, 2009; Wani et al., 2021). Furthermore, WRKY TFs also work in coordination with TGA TFs to regulate SA biosynthesis and perception via NPR1 (Chen et al., 2019). In addition, as mentioned in the above section, the genes EDS1 and PAD4 are also exclusively involved in the regulation of NPR1 signaling, which, in turn, activate SA signaling and associated defense response in plants exposed to HT stress (Feys et al., 2001; Rietz et al., 2011). Recent findings have confirmed that EDS1 acts upstream of NPR1, thus mediating the translocation of NPR1 to the nucleus by mediating crosstalk with JA and ethylene (Wang et al., 2020). Overall, NPR1-dependent SA signaling is an important pathway that regulates the transcription of several defence-responsive genes/proteins in plants. It also helps establish stress memory, enabling plants to adapt, improvise, and respond to repeated biotic/abiotic stress conditions (Phukan et al., 2016). Lastly, enhanced SA signaling improves plants' innate immune response, allowing them to withstand and combat stress anomalies by stimulating secondary metabolite production.

## 2.2. NPR1-independent signaling

A large body of literature has connoted NPR1 as a central player in SA perception and signaling in plants. However, evidence also suggests the existence of NPR1-independent SA signaling pathways (Zhang et al., 2020; Ding et al., 2018). These pathways not only contribute to plant defense but also regulate the expression of several important genes in growth and developmental pathways (Ding et al., 2018). Two such genes, NPR3 and NPR4, are essential components of NPR1-independent signaling and are activated in the absence of functional NPR1 (Fu et al., 2012; Vega et al., 2019). Several reports have conceded that both NPR3 and NPR4 are members of the Toll-interleukin 1 receptor-nucleotide-binding-leucine-rich repeat (TIR-NB-LRR) protein family localized in the cytoplasm (Li et al., 2004; Santolini et al., 2017). Like NPR1, NPR3, and NPR4 also act as SA receptors, leading to the activation of the SA signaling pathway (Song et al., 2008). The binding of SA to NPR3 and NPR4 occurs via the ligand binding domain (LBD) present in the N-terminal region of NPR3 and NPR4 (Wu et al., 2012; Chen et al., 2021). The LBD of NPR3 and NPR4 acts as a pocket where SA binds, and it is also believed that the accommodation of the SA molecule within this pocket interacts and activates NPR3 and NPR4 (Liu et al., 2022).

Upon interaction with the receptor, SA induces a conformational change in NPR3 and NPR4, thus rearranging specific domains or regions in the receptor, thus activating SA biosynthesis and signaling (Song et al., 2008). SA-mediated activation of NPR3 and NPR4 acts as co-activators and interacts with other receptors and cofactors to regulate downstream signaling events (Liu et al., 2020; Chamizo-Ampudia et al., 2017). Conceding the above notion, SA binding to NPR3 and NPR4 favors their interaction with TGA TF, which also stimulates SA-mediated defense responses in plants (Fu et al., 2012; Li et al., 2004). In addition, the interaction between NPR3/NPR4 and TGA TFs often leads to the sequestration of various defense-responsive genes/proteins, thus preventing the suppression of gene expression (Fu et al., 2012).

Another aspect of NPR1-independent SA signaling involves proteasomal degradation of unfolded proteins. Upon exposure of plants to biotic/abiotic stress and subsequent interaction with TFs, both NPR3 and NPR4 undergo ubiquitination for their degradation via the proteasome (Ding et al., 2018). These degradations unleash temporary inhibition from TFs, allowing them to bind their cis-regulatory elements and activate the expression of stress-responsive genes (Xu et al., 2022). Studies have reported that an abundance of NPR3 and NPR4 follows negative-feedback regulation, where excess SA biosynthesis leads to the degradation of NPR3 and NPR4 proteins, resulting in the activation of TFs (Flors et al., 2008). Cumulatively, NPR3 and NPR4 receptors regulate the SA signaling pathway without functional NPR1, thus instigating regulation in plant defence responses. Further, the interaction of NPR3/NPR4 with TFs and their subsequent degradation activates the transcription of stress-responsive genes, thus strengthening plant growth and immunity under HT stress. It is also conceded that future research regarding the precise structural information of the SA-NPR3/NPR4 complex and identifying interacting partners will help uncover the exact binding mode and the subsequent signaling events triggered by SA binding.

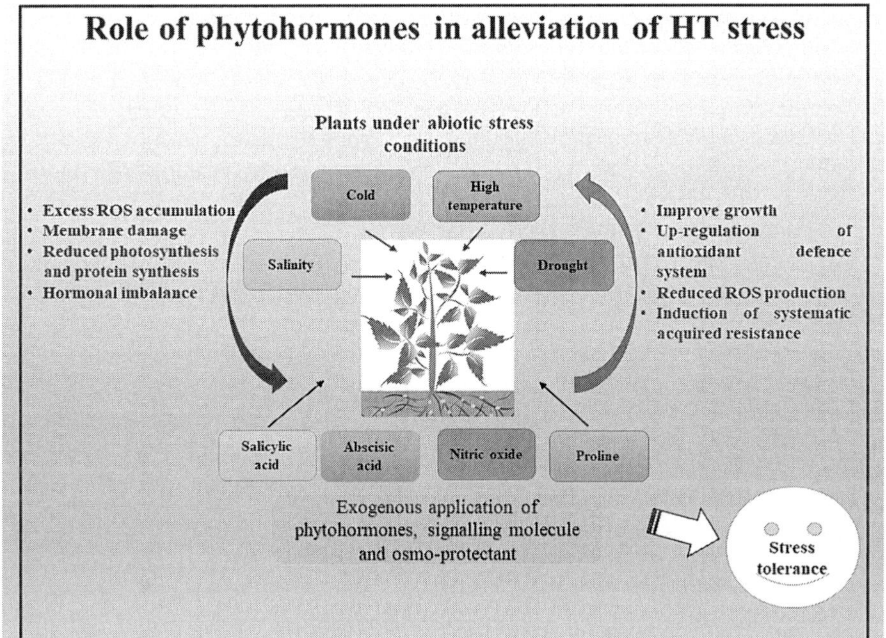


**Figure 1** An overview of plant phytohormones mediating stress signaling in plants.

## 3. SA perception

The perception and signaling of SA in plants involve an intricate network of receptors, signaling molecules, and downstream responses. Specific environmental cues such as HT stress trigger SA biosynthesis in plants, which is then perceived by receptor proteins present in the plasma membrane or in the cytoplasm of plant cells (Ding et al., 2018). In plants, several SA receptor proteins, such as NPR1, NPR3, and NPR4, are known to regulate SA perception and signaling, thereby activating stress defense responses (Fu et al., 2012). Increasing reports have demarcated NPR1 as a central regulator for SA signaling, which remains in an oligomeric state in the absence of SA signaling (Table 2). Upon SA perception, its translocation to the nucleus activates NPR1 activity by converting it into a monomer (Wu et al., 2012). Additionally, the activated NPR1 also interacts and activates TGA TFs, which cumulatively activate several downstream signaling genes (Pandey and Somssich, 2009).

The above transcriptional reprogramming of several genes stimulates the biosynthesis of important secondary metabolites, such as antimicrobial proteins, pathogenesis-related (PR) proteins, and several enzymatic and non-enzymatic antioxidants (Lastochkin et al., 2022). SAR is a phenomenon that is established upon exaggerated SA signaling and also stimulates SA perception by activating SA receptor proteins (Fu and Dong, 2013). Furthermore, SA perception is also modulated via feedback regulation, which is exclusively required for maintaining SA homeostasis in plants experiencing HT stress (Chamizo-Ampudia et al., 2017). Under non-stimulative environments, the receptor proteins NPR3/NPR4 may function as a negative regulator of SA signaling by preventing the interaction of NPR1 with SA (Vega et al., 2019). This feedback regulation ensures a balanced supply of SA to the cells and prevents excessive activation of SA by stimulating the necessary degradation of NPR1 receptor protein (Ding et al., 2018).

Studies have also connotated the involvement of other phytohormones, such as jasmonic acid (JA) and ethylene (ET), in modulating SA signaling and perception (Pieterse et al., 2012; Van der Does et al., 2013). Increasing crosstalk of SA with other phytohormones, perse., JA and ET often intersect at the level of signaling and perception, which is believed to be modulated by shared signaling components (Panstruga and Kuhn, 2019). An increasing body of literature has pointed to the antagonistic role of JA and ET in modulating SA signaling and perception (Liu et al., 2019). Under pathogenic attack, both JA and ET signaling are dynamically regulated, which actively modulates the degradation of NPR1, leading to the suppression of SA signaling (Pieterse et al., 2012; Van der Does et al., 2013). SA signaling and perception have been known to be primarily associated with biotrophic pathogens, whereas JA/ET signaling is actively related to necrotrophic pathogens (Pieterse et al., 2012; Kazan and Manners, 2013). Interestingly, researchers have also conceded the synergistic interaction of SA with JA/ET, where enhanced JA/ET signaling has enhanced SA signaling and perception, leading to the activation of defense responses in plants (Pieterse et al., 2012). The exact mechanism of how JA/ET act synergistically with SA is poorly understood. Yet, scientists have connotated that specific environmental factors such as HT, highlight, and pathogens could shift the above momentum, favouring SA/JA/ET to exert a defence response in plants.

## 4. Nitrite, a significant NO signaling component in plants

Similar to SA, NO signaling and perception also involve complex signaling pathways that regulate plant growth and defense under stress conditions (Gibbs et al., 2014; Qiao and Fan, 2008). Like animals, plants do not possess nitric oxide synthase (NOS) enzyme, which catalyzes the NO production in animals (Gupta et al., 2011). Increasing literature has confirmed that plants produce NO through two distinct pathways, i.e., enzymatic and non-enzymatic. However, one of the major pathways for NO production in plants involves the nitrate-nitrite-NO pathway (Vega et al., 2019; Lau et al., 2021). The nitrate-nitrite-NO pathway in plants encompasses a series of enzymatic reactions catalyzed by the nitrate reductase (NR) enzyme (Gupta et al., 2011; Qiao and Fan, 2008). As the plant absorbs nitrate from the soil through specific transporters present in root cell cytoplasm, it is then transported to the leaves for further reduction (Moreau et al., 2010; Lau et al., 2021). The NR enzyme remains inactive in plant cells, and to catalyze nitrate reduction to NO, it is activated by an NR-activating enzyme that catalyzes the reduction of a disulfide bond in NR (Gibbs et al., 2014; Qiao and Fan, 2008).

After its activation, NR catalyzes the reduction of nitrate to NO by using NADPH as a cofactor, which transfers an electron to stimulate the reduction reaction (Rockel et al., 2002; Mallikarjuna et al., 2022). Furthermore, in the above response, NADPH transfers two electrons to nitrate, resulting in the generation of NO via a multi-step reaction using several redox cofactors (Qiao and Fan, 2008; Gupta et al., 2020). In *Arabidopsis thaliana*, the NR gene family consists of NIA1, NIA2, and NIA3 genes, where NIA1 and NIA2 have been reported to encode cytosolic NR isoforms, and NIA3 encodes a mitochondrial NR isoform (Liu et al., 2019; Gupta et al., 2020). Similarly, the enzyme nitrite reductase (NiR) also plays a crucial role in catalyzing the signaling of NO by converting nitrate to nitrite and finally to NO (Gupta et al., 2011; Lau et al., 2021). Like NR, two isoforms of NiR, i.e., NiR1 and NiR2, have also been reported in Arabidopsis (Gupta et al., 2020).

**Table 2** Physiological, biochemical, and molecular mechanisms of perception of heat stress by salicylic acid (SA) and nitric oxide (NO) in plants

| Mechanism | Physiological Responses | Biochemical Responses | Molecular Responses | References |
|---|---|---|---|---|
| **Perception by Stomatal Closure** | SA and NO regulate stomatal closure under heat stress conditions. | Stomatal closure reduces transpiration rates and prevents excessive water loss. | SA induces stomatal closure through the activation of guard cell ion channels and signaling cascades involving ROS and $Ca^{2+}$. NO acts synergistically with SA to promote stomatal closure by modulating ion transporters and regulating guard cell signaling. | Zhang et al. (2013); Wang et al. (2019). |
| **Heat Shock Protein (HSP) Induction** | SA and NO regulate the induction of HSPs under heat stress. | HSPs assist in protein folding and prevent protein aggregation under heat stress. | SA and NO signaling pathways activate HSF transcription factors, which bind to HSE in the promoters of HSP genes, resulting in their upregulation. | Mishra et al. (2013); Zinta et al. (2022) |

| | | | | |
|---|---|---|---|---|
| **Antioxidant Defense Activation** | SA and NO regulate the activation of antioxidant defense mechanisms under heat stress. | Antioxidant enzymes scavenge ROS and protect cellular components from oxidative damage. | SA and NO signaling pathways upregulate the expression and activity of antioxidant enzymes, and induce the synthesis of non-enzymatic antioxidants. | Hasanuzzaman et al. (2013); Qiao and Fan, 2008 |
| **Transcriptional Regulation** | SA and NO modulate the expression of heat stress-responsive genes through transcriptional regulation. | Heat stress-responsive genes play crucial roles in plant adaptive responses to heat stress. | SA and NO signaling pathways activate specific transcription factors, resulting in the upregulation of heat stress-responsive genes. | Suzuki et al. (2016); Lau et al. (2021) |

NAD(P)H-dependent nitrite reductase genes are also involved in regulating the signaling and perception of NO in plants (Rather, 2020). Several NAD(P)H-dependent nitrite reductase genes, such as NIN-like proteins (NLP) genes, including NLP3, NLP4, and NLP7, have been identified and characterized in plants, playing unique roles in donating electrons for NO signaling (Gupta et al., 2020; Rather, 2020). Studies have also mentioned the existence of hemoglobin-dependent NO production in plants by stimulating the reduction of nitrite to NO (Astier et al., 2018; Zeier et al., 2004). The mechanism behind this notion involves the binding of nitrite to the heme iron center of hemoglobin, which stimulates other genes involved in catalyzing the reduction of nitrite to NO (Zeier et al., 2004; Chamizo-Ampudia et al., 2017). Genes such as GLB1, GLB2, GLB3, and GLB4 have been identified from Arabidopsis that stimulate hemoglobin-dependent NO signaling in plants (Zeier et al., 2004; Chamizo-Ampudia et al., 2017).

Several nitrate transporter genes have also been reported to modulate NO signaling and perception in plants by enhancing the availability of nitrate to plants (Vega et al., 2019; Santolini et al., 2017). In Arabidopsis, NRT1 and NRT2 genes have been known to stimulate the above process, of which NRT2.1, NRT2.2, and NRT2.4 robustly function to stimulate nitrate uptake (Krapp et al., 2024; Krouk et al., 2006). The non-enzymatic antioxidant pathway genes also act responsibly to regulate NO signaling and perception in plants (Mallikarjuna et al., 2022). Ascorbate biosynthetic genes are actively involved in the stimulation of the non-enzymatic reduction of nitrite to NO (Tillmann, 2021). L-galactose-1-phosphate phosphatase (VTC2) and L-galactono-1,4-lactone dehydrogenase (VTC4) genes have been identified in Arabidopsis that stimulate ascorbate biosynthesis and are also exceptionally involved in NO signaling and perception in plants (Arasimowicz and Floryszak-Wieczorek, 2007; Santolini et al., 2017).

In summary, the nitrate-nitrite-NO pathway catalyses the enzymatic reduction of nitrate to nitrite by nitrate reductase, which is then converted to nitrite and NO via enzymatic (nitrite reductase) and non-enzymatic (ascorbic acid, thiols) mechanisms. In addition, haemoglobin-dependent NO production is also significantly involved in the above pathway, especially in those plants that possess haemoglobins. Once NO is generated, its perception through specific receptors enables it to function as a signaling molecule, thus regulating many physiological and biochemical processes in plants. It is essential to mention that the NR-mediated reduction of nitrate to nitrite is a critical step in the nitrate-nitrite-NO pathway. Further reduction of nitrite to NO through various enzymatic and non-enzymatic mechanisms mentioned above are essential in modulating NO signaling. Nonetheless, it's worth noting that the mechanism of nitrate reduction, especially haemoglobin and non-enzymatic reduction, can vary depending on plant species and geography, where certain environmental conditions could be the critical regulator of the mechanisms. However, in most plants, both mechanisms remain constant and effective for regulating NO signaling and perception under stress conditions.

## 5. A NOS-like enzyme in regulating NO signaling

Nitric oxide synthase (NOS) is an enzyme that catalyzes the production of NO in animals (Astier et al., 2018). Recent discoveries have unraveled the existence of NOS-like activity in plants, which is involved in the biosynthesis of NO from L-arginine (Gibbs et al., 2014). In Arabidopsis thaliana, researchers have identified one such NOS-like enzyme, i.e., NO-Associated Protein1 (AtNOA1), which mediates the biosynthesis of NO from L-arginine (Yun et al., 2011). Further, NOA1 has also been identified as a vital modulator of NO signaling and perception, thus regulating various physiological processes in plants exposed to HT stress (Tillmann, 2021). Additionally, NOA1 also actively maintains NO homeostasis in plants by interacting with another signaling molecule as well as activating genes and proteins involved in the nitrate-nitrite-NO pathway (Liu et al., 2019). Increasing literature has also suggested its intricate role in regulating growth and developmental processes in plants (Table 2), ranging from seed germination to root growth and flowering/fruiting initiation (Santolini et al., 2017; Fu et al., 2012).

Recent investigations have identified other NOS-like proteins (NLP) in Arabidopsis, i.e., AtNOS1/NOS1-like proteins that share exquisite similarity with the animal NOS enzyme (Sarwat and Tuteja, 2018). Similarly, studies have also unraveled the existence of NOS-like protein (OsNOS1/NLP) in *Oryza sativa*, which has been shown to stimulate NO signaling and perception (Gupta et al., 2020). Additionally, researchers have also identified various proteins in distinct plant species, such as Glycine max and Nicotiana tabacum, that share extensive homology with NOS-like proteins. However, their role in NO signaling and perception is still under investigation (Santolini et al., 2017; del Rıo et al., 2004).

Among all the NOS-like proteins identified in plants, the most extensively studied is the AtNOS1/NOS1-like protein of Arabidopsis (Moreau et al., 2010). In-depth investigations have confirmed its role in the production, perception, and signaling of NO in plants via a mechanism analogous to that of NOS-like protein found in animals. One of the features which are common in both plant and animal NOS-like proteins is the presence of oxygenase and reductase domains that are responsible for their catalytic activity (Lozano-Juste and León, 2011). It is because of this feature that plant scientists are convinced that AtNOS1/NOS1-like proteins play a role in NO biosynthesis (Chen et al., 2021). To further validate the part of AtNOS1/NOS1-like protein in NO biosynthesis, researchers performed an in-vitro enzymatic activity assay. Their results indicated that AtNOS1/NOS1-like protein catalyzes the conversion of L-arginine to L-citrulline, which is an essential prerequisite for NOS enzymes for stimulating the biosynthesis of NO in plants (Corpas et al., 2022).

Correspondingly, recent findings have also conceded that the activity of AtNOS1/NOS1-like proteins in plants is also modulated by environmental cues and interaction with other signaling molecules and phytohormones (Zinta et al., 2022). For instance, calcium/calmodulin ($Ca^{2+}$/CaM) is a potential modulator of NOS1-like proteins in plants. Under specific conditions, such as biotic/abiotic stresses, these $Ca^{2+}$/CaM interact and activate NOS1-like proteins due to an exaggerated influx of $Ca^{2+}$, which may trigger regulatory components to activate NOS enzyme (Gao et al., 2015). Some studies have also suggested that NOS-like proteins interact with specific protein kinases to modulate their regulatory functions (Fu et al., 2021). For example, receptor-like kinases (RLKs) or calcium-dependent protein kinases (CDPKs) modulate the biosynthesis and signaling of NO in plants by activating NOS-like protein, either by stimulating downstream signaling genes or other signaling molecules, such as free oxygen radicals (Zinta et al., 2022). However, the comprehensive understanding of localisation and mechanisms of NOS-like proteins in plants is still vague to date. Therefore, extensive efforts are required to map NOS-like protein's precise localisation and regulatory component.

## Conclusion

SA and NO signaling in plants are imperative for improving their growth and development under heat stress responses. SA and NO signaling are explicitly important for regulating physiological, biochemical and molecular mechanisms, such as enzymatic and non-enzymatic antioxidants and stress-responsive genes/proteins. The interaction/crosstalk between SA and NO renders plants' complex network of heat stress responses, enabling them to improvise and survive under HT stress. In addition, the interplay between SA and NO signaling and perception rewires plant heat stress responses, ultimately enhancing plant resilience and survival under HT stress. However, future studies must be directed by exploiting OMICS techniques in conjunction with other system biology tools, such as yeast two-hybrid screens or co-immunoprecipitation, to gain valuable insight into the regulatory role of NOS-like protein in stimulating NO biosynthesis and signaling in plants. Further, studies directed towards elucidation of localisation of NOS-like protein will help reveal interaction partners of AtNOS1/NLP, unleashing underlying mechanisms of NO-related processes in Arabidopsis and potentially other plant species.

## References

Aarts, N., Metz, M., Holub, E., Staskawicz, B.J., Daniels, M.J. and Parker, J.E., 1998. Different requirements for EDS1 and NDR1 by disease resistance genes define at least two R gene-mediated signaling pathways in Arabidopsis. Proceedings of the National Academy of Sciences, 95(17), pp.10306-10311.

Arasimowicz, M. and Floryszak-Wieczorek, J., 2007. Nitric oxide as a bioactive signaling molecule in plant stress responses. Plant science, 172(5), pp.876-887.

Arasimowicz-Jelonek, M. and Floryszak-Wieczorek, J., 2014. Nitric oxide: an effective weapon of the plant or the pathogen?. Molecular Plant Pathology, 15(4), pp.406-416.

Astier, J. and Lindermayr, C., 2012. Nitric oxide-dependent posttranslational modification in plants: an update. International journal of molecular sciences, 13(11), pp.15193-15208.

Astier, J., Gross, I. and Durner, J., 2018. Nitric oxide production in plants: an update. Journal of Experimental Botany, 69(14), pp.3401-3411.

Aygören, A.S., Güneş, E., Muslu, S., Kasapoğlu, A.G., Yiğider, E., Aydın, M., Büyük, İ. and İlhan, E., 2023. Genome-wide analysis and characterization of SABATH gene family in Phaseolus vulgaris genotypes subject to melatonin under drought and salinity stresses. Plant Molecular Biology Reporter, 41(2), pp.242-259.

Chamizo-Ampudia, A., Sanz-Luque, E., Llamas, A., Galvan, A. and Fernandez, E., 2017. Nitrate reductase regulates plant nitric oxide homeostasis. Trends in Plant Science, 22(2), pp.163-174.

Chen, J., Zhang, J., Kong, M., Freeman, A., Chen, H. and Liu, F., 2021. More stories to tell: NONEXPRESSOR OF PATHOGENESIS-RELATED GENES1, a salicylic acid receptor. Plant, Cell & Environment, 44(6), pp.1716-1727.

Chen, X., Li, C., Wang, H. and Guo, Z., 2019. WRKY transcription factors: evolution, binding, and action. Phytopathology Research, 1(1), pp.1-15.

Coqueiro, D.S.O., de Souza, A.A., Takita, M.A., Rodrigues, C.M., Kishi, L.T. and Machado, M.A., 2015. Transcriptional profile of sweet orange in response to chitosan and salicylic acid. BMC genomics, 16, pp.1-14.

Corpas, F.J., González-Gordo, S. and Palma, J.M., 2022. NO source in higher plants: present and future of an unresolved question. Trends in Plant Science, 27(2), pp.116-119.

Cui, H., Gobbato, E., Kracher, B., Qiu, J., Bautor, J. and Parker, J.E., 2017. A core function of EDS1 with PAD4 is to protect the salicylic acid defense sector in Arabidopsis immunity. New Phytologist, 213(4), pp.1802-1817.

Cui, H., Tsuda, K. and Parker, J.E., 2015. Effector-triggered immunity: from pathogen perception to robust defense. Annual review of plant biology, 66, pp.487-511.

del Rıo, L.A., Corpas, F.J. and Barroso, J.B., 2004. Nitric oxide and nitric oxide synthase activity in plants. Phytochemistry, 65(7), pp.783-792.

Dempsey, D.M.A., Vlot, A.C., Wildermuth, M.C. and Klessig, D.F., 2011. Salicylic acid biosynthesis and metabolism. The Arabidopsis book/American Society of Plant Biologists, 9.

Ding, Y., Sun, T., Ao, K., Peng, Y., Zhang, Y., Li, X. and Zhang, Y., 2018. Opposite roles of salicylic acid receptors NPR1 and NPR3/NPR4 in transcriptional regulation of plant immunity. Cell, 173(6), pp.1454-1467.

Feys, B.J., Moisan, L.J., Newman, M.A. and Parker, J.E., 2001. Direct interaction between the Arabidopsis disease resistance signaling proteins, EDS1 and PAD4. The EMBO journal, 20(19), pp.5400-5411.

Flors, V., Ton, J., Van Doorn, R., Jakab, G., García-Agustín, P. and Mauch-Mani, B., 2008. Interplay between JA, SA and ABA signaling during basal and induced resistance against Pseudomonas syringae and Alternaria brassicicola. The Plant Journal, 54(1), pp.81-92.

Fu, Z.Q. and Dong, X., 2013. Systemic acquired resistance: turning local infection into global defense. Annual review of plant biology, 64, pp.839-863.

Fu, Z.Q., Yan, S., Saleh, A., Wang, W., Ruble, J., Oka, N., Mohan, R., Spoel, S.H., Tada, Y., Zheng, N. and Dong, X., 2012. NPR3 and NPR4 are receptors for the immune signal salicylic acid in plants. Nature, 486(7402), pp.228-232.

Gao, Y., Zhang, Y., Zhang, D., Dai, X., Estelle, M. and Zhao, Y., 2015. Auxin binding protein 1 (ABP1) is not required for either auxin signaling or Arabidopsis development. Proceedings of the National Academy of Sciences, 112(7), pp.2275-2280.

Germain, H., Qu, N., Cheng, Y.T., Lee, E., Huang, Y., Dong, O.X., Gannon, P., Huang, S., Ding, P., Li, Y. and Sack, F., 2010. MOS11: a new component in the mRNA export pathway. PLoS genetics, 6(12), p.e1001250.

Gibbs, D.J., Isa, N.M., Movahedi, M., Lozano-Juste, J., Mendiondo, G.M., Berckhan, S., Marin-de La Rosa, N., Conde, J.V., Correia, C.S., Pearce, S.P. and Bassel, G.W., 2014. Nitric oxide sensing in plants is mediated by proteolytic control of group VII ERF transcription factors. Molecular Cell, 53(3), pp.369-379.

Gupta, K.J., Fernie, A.R., Kaiser, W.M. and van Dongen, J.T., 2011. On the origins of nitric oxide. Trends in plant science, 16(3), pp.160-168.

Gupta, K.J., Hancock, J.T., Petrivalsky, M., Kolbert, Z., Lindermayr, C., Durner, J., Barroso, J.B., Palma, J.M., Brouquisse, R., Wendehenne, D. and Corpas, F.J., 2020. Recommendations on terminology and experimental best practice associated with plant nitric oxide research. New Phytologist, 225(5), pp.1828-1834.

Kazan, K. and Manners, J.M., 2013. MYC2: the master in action. Molecular plant, 6(3), pp.686-703.

Keech, O., Gardeström, P., Kleczkowski, L.A. and Rouhier, N., 2017. The redox control of photorespiration: from biochemical and physiological aspects to biotechnological considerations. Plant, cell & environment, 40(4), pp.553-569.

Keszler, A., Zhang, Y. and Hogg, N., 2010. Reaction between nitric oxide, glutathione, and oxygen in the presence and absence of protein: How are S-nitrosothiols formed?. Free Radical Biology and Medicine, 48(1), pp.55-64.

Krapp, A., David, L.C., Chardin, C., Girin, T., Marmagne, A., Leprince, A.S., Chaillou, S., Ferrario-Méry, S., Meyer, C. and Daniel-Vedele, F., 2014. Nitrate transport and signaling in Arabidopsis. Journal of experimental botany, 65(3), pp.789-798.

Krouk, G., Tillard, P. and Gojon, A., 2006. Regulation of the high-affinity NO3− uptake system by NRT1. 1-mediated NO3− demand signaling in Arabidopsis. Plant Physiology, 142(3), p.1075.

Lastochkina, O., Aliniaeifard, S., SeifiKalhor, M., Bosacchi, M., Maslennikova, D. and Lubyanova, A., 2022. Novel approaches for sustainable horticultural crop production: Advances and prospects. Horticulturae, 8(10), p.910.

Lau, S.E., Hamdan, M.F., Pua, T.L., Saidi, N.B. and Tan, B.C., 2021. Plant nitric oxide signaling under drought stress. Plants, 10(2), p.360.

Li, J., Brader, G. and Palva, E.T., 2004. The WRKY70 transcription factor: a node of convergence for jasmonate-mediated and salicylate-mediated signals in plant defense. The Plant Cell, 16(2), pp.319-331.

Lorrain, S., Vailleau, F., Balagué, C. and Roby, D., 2003. Lesion mimic mutants: keys for deciphering cell death and defense pathways in plants?. Trends in plant science, 8(6), pp.263-271.

Lozano-Juste, J. and León, J., 2011. Nitric oxide regulates DELLA content and PIF expression to promote photomorphogenesis in Arabidopsis. Plant Physiology, 156(3), pp.1410-1423.

Mallikarjuna, B.P., Shettigar, N., Radhika, D.H., Devi, E.L., Bhat, J.S., Patil, B.S., Lohithaswa, H.C. and Mallikarjuna, M.G., 2022. Genome-

Wide Association Studies and Genomic Selection for Nutrient Use Efficiency in Cereals. In Next-Generation Plant Breeding Approaches for Stress Resilience in Cereal Crops (pp. 161-197). Singapore: Springer Nature Singapore.

Mauch-Mani, B. and Mauch, F., 2005. The role of abscisic acid in plant–pathogen interactions. Current opinion in plant biology, 8(4), pp.409-414.

Mishina, T.E. and Zeier, J., 2006. The Arabidopsis flavin-dependent monooxygenase FMO1 is an essential component of biologically induced systemic acquired resistance. Plant physiology, 141(4), pp.1666-1675.

Moreau, M., Lindermayr, C., Durner, J. and Klessig, D.F., 2010. NO synthesis and signaling in plants–where do we stand?. Physiologia Plantarum, 138(4), pp.372-383.

Mou, Z., Fan, W. and Dong, X., 2003. Inducers of plant systemic acquired resistance regulate NPR1 function through redox changes. Cell, 113(7), pp.935-944.

Pandey, S.P. and Somssich, I.E., 2009. The role of WRKY transcription factors in plant immunity. Plant physiology, 150(4), pp.1648-1655.

Panstruga, R. and Kuhn, H., 2019. Mutual interplay between phytopathogenic powdery mildew fungi and other microorganisms. Molecular Plant Pathology, 20(4), pp.463-470.

Phukan, U.J., Jeena, G.S. and Shukla, R.K., 2016. WRKY transcription factors: molecular regulation and stress responses in plants. Frontiers in plant science, 7, p.760.

Pieterse, C.M., Van der Does, D., Zamioudis, C., Leon-Reyes, A. and Van Wees, S.C., 2012. Hormonal modulation of plant immunity. Annual review of cell and developmental biology, 28, pp.489-521.

Qiao, W. and Fan, L.M., 2008. Nitric oxide signaling in plant responses to abiotic stresses. Journal of Integrative Plant Biology, 50(10), pp.1238-1246.

Rather, B.A., Mir, I.R., Sehar, Z., Anjum, N.A., Masood, A. and Khan, N.A., 2020. The outcomes of the functional interplay of nitric oxide and hydrogen sulfide in metal stress tolerance in plants. Plant Physiology and Biochemistry, 155, pp.523-534.

Rietz, S., Stamm, A., Malonek, S., Wagner, S., Becker, D., Medina-Escobar, N., Corina Vlot, A., Feys, B.J., Niefind, K. and Parker, J.E., 2011. Different roles of Enhanced Disease Susceptibility1 (EDS1) bound to and dissociated from Phytoalexin Deficient4 (PAD4) in Arabidopsis immunity. New Phytologist, 191(1), pp.107-119.

Rockel, P., Strube, F., Rockel, A., Wildt, J. and Kaiser, W.M., 2002. Regulation of nitric oxide (NO) production by plant nitrate reductase in vivo and in vitro. Journal of experimental botany, 53(366), pp.103-110.

Santolini, J., André, F., Jeandroz, S. and Wendehenne, D., 2017. Nitric oxide synthase in plants: where do we stand?. Nitric Oxide, 63, pp.30-38.

Sarwat, M. and Tuteja, N. eds., 2018. Senescence signaling and control in plants. Academic Press.

Song, L., Ding, W., Shen, J., Zhang, Z., Bi, Y. and Zhang, L., 2008. Nitric oxide mediates abscisic acid induced thermotolerance in the calluses from two ecotypes of reed under heat stress. Plant Science, 175(6), pp.826-832.

Song, X., Wang, T., Zhang, Y., Yu, J.Q. and Xia, X.J., 2022. S-Nitrosoglutathione Reductase Contributes to Thermotolerance by Modulating High Temperature-Induced Apoplastic H2O2 in Solanum lycopersicum. Frontiers in Plant Science, 13, p.862649.

Su, X., Qi, X. and Cheng, H., 2014. Molecular cloning and characterization of enhanced disease susceptibility 1 (EDS1) from Gossypium barbadense. Molecular biology reports, 41, pp.3821-3828.

Tillmann, M.A., 2021. The Auxin Biosynthesis Metabolic Network (Doctoral dissertation, University of Minnesota).

Van der Does, D., Leon-Reyes, A., Koornneef, A., Van Verk, M.C., Rodenburg, N., Pauwels, L., Goossens, A., Körbes, A.P., Memelink, J., Ritsema, T. and Van Wees, S.C., 2013. Salicylic acid suppresses jasmonic acid signaling downstream of SCFCOI1-JAZ by targeting GCC promoter motifs via transcription factor ORA59. The Plant Cell, 25(2), pp.744-761.

Vega, A., O'Brien, J.A. and Gutiérrez, R.A., 2019. Nitrate and hormonal signaling crosstalk for plant growth and development. Current Opinion in Plant Biology, 52, pp.155-163.

Vlot, A.C., Dempsey, D.M.A. and Klessig, D.F., 2009. Salicylic acid, a multifaceted hormone to combat disease. Annual review of phytopathology, 47, pp.177-206.

Wang, D., Amornsiripanitch, N. and Dong, X., 2006. A genomic approach to identify regulatory nodes in the transcriptional network of systemic acquired resistance in plants. PLoS pathogens, 2(11), p.e123.

Wang, Y.Y., Cheng, Y.H., Chen, K.E. and Tsay, Y.F., 2018. Nitrate transport, signaling, and use efficiency. Annual review of plant biology, 69, pp.85-122.

Wani, S.H., Anand, S., Singh, B., Bohra, A. and Joshi, R., 2021. WRKY transcription factors and plant defense responses: Latest discoveries and future prospects. Plant Cell Reports, 40, pp.1071-1085.

Waterhouse, P.M., Graham, M.W. and Wang, M.B., 1998. Virus resistance and gene silencing in plants can be induced by simultaneous expression of sense and antisense RNA. Proceedings of the National Academy of Sciences, 95(23), pp.13959-13964.

Wiermer, M., Feys, B.J. and Parker, J.E., 2005. Plant immunity: the EDS1 regulatory node. Current opinion in plant biology, 8(4), pp.383-389.

Wu, Y., Zhang, D., Chu, J.Y., Boyle, P., Wang, Y., Brindle, I.D., De Luca, V. and Després, C., 2012. The Arabidopsis NPR1 protein is a receptor for the plant defense hormone salicylic acid. Cell reports, 1(6), pp.639-647.

Xu, J., Zhao, X., Bao, J., Shan, Y., Zhang, M., Shen, Y., Abubakar, Y.S., Lu, G., Wang, Z. and Wang, A., 2022. Genome-wide identification and expression analysis of SNARE genes in Brassica napus. Plants, 11(5), p.711.

Yun, B.W., Feechan, A., Yin, M., Saidi, N.B., Le Bihan, T., Yu, M., Moore, J.W., Kang, J.G., Kwon, E., Spoel, S.H. and Pallas, J.A., 2011. S-nitrosylation of NADPH oxidase regulates cell death in plant immunity. Nature, 478(7368), pp.264-268.

Zeier, J., Delledonne, M., Mishina, T., Severi, E., Sonoda, M. and Lamb, C., 2004. Genetic elucidation of nitric oxide signaling in incompatible plant-pathogen interactions. Plant Physiology, 136(1), pp.2875-2886.

Zhang, X., Yang, Z., Wu, D. and Yu, F., 2020. RALF–FERONIA signaling: linking plant immune response with cell growth. Plant communications, 1(4).

Zinta, G., Singh, R.K. and Kumar, R., 2022. Cold adaptation strategies in plants—An emerging role of epigenetics and antifreeze proteins to engineer cold resilient plants. Frontiers in Genetics, 13, p.909007.

# CHAPTER 6

# SALICYLIC ACID AND NITRIC OXIDE: TRANSCRIPTION REGULATORY NETWORK UNDER HEAT STRESS

## Abstract

Heat stress is a significant environmental challenge that adversely affects plant growth and development, reducing crop productivity. Plants employ intricate signaling networks to perceive and respond to heat stress. Among these signaling pathways, salicylic acid (SA) and nitric oxide (NO) have emerged as key players in regulating plant responses to heat stress. The interplay between SA and NO signaling pathways under heat stress further strengthens the transcription regulatory network. SA and NO signaling can synergistically activate transcription factors and coordinately modulate gene expression. They interact with other signaling molecules and pathways, including reactive oxygen species (ROS) and phytohormones, to fine-tune heat stress responses. Further research in this area will provide deeper insights into the specific interactions, downstream targets, and regulatory pathways involved in the SA and NO-mediated transcriptional responses to heat stress. Such knowledge can contribute to developing strategies to enhance plant thermotolerance and improve crop productivity under heat-stress conditions.

**Keywords**: Salicylic acid, Nitric oxide, Transcription regulatory network, Heat stress, Plant responses

## Introduction

The transcriptional regulatory network of SA and NO under heat stress involves the activation of distinct signaling pathways and molecules to coordinate the expression of various transcription factors and stress-responsive genes/proteins (Rivas-San Vicente et al., 2013; Meng et al., 2022; Sharma et al., 2013). SA is a plant hormone, whereas NO is a gaseous signaling molecule known to play an unprecedented role in stimulating plant innate immune response under various stress conditions, including heat stress (Iqbal et al., 2019; Rai et al., 2020; Hu et al., 2006). SA-mediated induction of the transcriptional regulatory network is initiated upon the biosynthesis of SA either by the isochorismate pathway or the phenylalanine ammonia-lyase (PAL) pathway (Rivas-San Vicente et al., 2013; Bellin et al., 2013; Sharma et al., 2013). Following synthesis, the perception of SA through NPR1 (oligomeric) or TGA occurs within the cytoplasm leading to the activation of NPR1 (monomeric) upon its translocation to the nucleus (Zhang et al., 2012; Zhou et al., 2000; Rai et al., 2020).

The activated NPR1 within the nucleus interacts with nuclear-localized SA receptors to activate the expression of transcription factors (TFs) and other stress-responsive genes (Zhang et al., 2012; Spoel et al., 2019; Pieterse et al., 2012). Among various TFs, WRKY TFs are the first responders and regulators of SA-mediated heat stress responses in plants (Li et al., 2017; Qiao and Fan, 2008; Phukan et al., 2016). Apart from WRKY, other TFs such as Apetella 2 and Ethylene Responsive factor (AP2/ERF), No apical meristem, Arabidopsis transcription activator factor, and Cup-shaped cotyledon (NAC), dehydration-responsive element-binding protein (DREB), basic leucine zipper (bZIP), and myeloblastosis (MYB) are also involved in modulating plant defense responses under stress conditions (Nakashima et al., 2014; Bhargava and Sawant, 2013; He et al., 2012).

Similarly, NO-mediated transcriptional reprogramming is also initiated upon enzymatic (Nitrate-Nitrite pathway) and non-enzymatic (S-nitrosothiol) pathways for NO production (Rather et al., 2020; Mur et al., 2013; McCarthy-Suárez, 2017). Once produced, NO perception by various receptor proteins and cofactors activates defense signaling pathways (Rather et al., 2020; Salmi et al., 2022). Like SA, NO also induces the transcriptional regulation of various TFs such as NAC, DREB, AP2/ERF, and MYB, which in turn activate the expression of other stress-responsive genes/proteins (Yin et al., 2013; Qiao and Fan, 2008; Wei et al., 2017). It is worth noting that, like SA, NO signaling under heat stress maintains and regulates cellular homeostasis in plants by modulating the expression of heat shock proteins (HSPs) (Paul and Roychoudhury, 2020; Shan et al., 2021; Zandalinas et al., 2022). The HSPs, particularly HSP70 and HSP90, have been reported to be most commonly induced by NO signaling and perception to alleviate heat stress-induced oxidative damage in plants (Hu et al., 2006; Zhou et al., 2000; Shan et al., 2021).

SA and NO also modulate transcriptional reprogramming by mediating the cross-talk with other phytohormones such as jasmonic acid (JA), ethylene (ET), and abscisic acid (ABA) (Alamri et al., 2019; Asgher et al., 2017; Lozano-Juste and León, 2011). Studies have reported that the interaction between SA and NO and among different phytohormones may exert synergistic or antagonistic responses on plant growth and developmental conditions under HT conditions (Jin et al., 2020; Astier et al., 2012; Gilroy and Breen, 2022). Nonetheless, the interplay of SA and NO with other plant growth regulators has been implicated in fine-tuning the plant defense response by modulating multiple stress signaling pathways and associated components such as secondary metabolite production (Lamattina et al., 2003; Hu et al., 2021).

## SA and NO: Transcriptional factor-mediated regulation

In plants, the perception and signaling of SA through NPR1 protein act as a central activator of TFs-mediated regulation of plant defense response under stress conditions (Fu et al., 2012; Wu et al., 2020; Fu and Dong, 2013; Pieterse et al., 2012). The dramatic increase in the level of SA provokes conformational activation of NPR1 in the nucleus, where it interacts and activates TGA TFs to induce the expression of downstream signaling genes (Spoel et al., 2009; Ding et al., 2023; McCarthy-Suárez, 2017). Similarly, NO is known to modulate the activity of TFs by interacting and regulating the activity of various signaling molecules such as kinases, phosphatases, and other regulatory proteins (Huang et al., 2016; Qiao and Fan, 2008; Ma et al., 2022). Another mechanism by which NO can activate TFs is by performing nitrosylation and tyrosine nitration of TFs, thus adding specific cysteine residues with TF proteins (Lindermayr et al., 2010; Mur et al., 2013; Parray et al., 2019). This section will provide an in-depth understanding of the functional mechanism of SA and NO in stimulating the expression of different TFs under HT stress (Table 1 and 2; Figure 1).

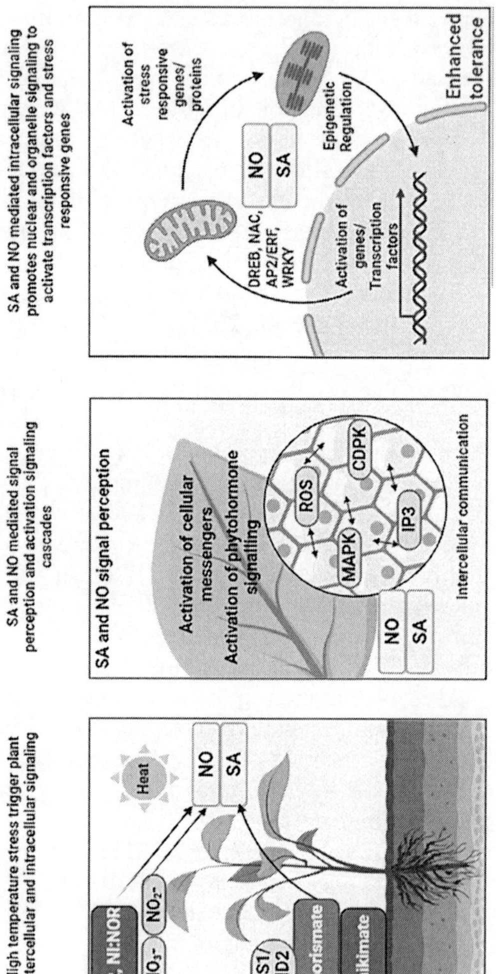


**Figure 1** An overview of functional mechanism of SA and NO in stimulating the expression of different TFs under HT stress.

## AP2/ERF

APETALA2/ethylene-responsive factor (AP2/ERF) is a member of one of the most prominent transcription factor families present in plants (Licausi et al., 2013; Song et al., 2020; Wei et al., 2017). AP2/ERF plays a specific role in regulating diverse plant functions, including growth and developmental responses under various environmental stresses (Yang et al., 2019; Wei et al., 2021; Hu et al., 2006). Studies have also implicated the role of AP2/ERF in improving plant defense against pathogens, insects/pests, and herbivores (Diao et al., 2020; Rai et al., 2020; Qiao et al., 2022). The notion behind this act of AP2/ERF is that once activated, these TFs regulate the expression of defense-responsive genes, particularly those encoding pathogenesis-related (PR) proteins (Hu et al., 2016; Zhou et al., 2000). The PR proteins, in turn, activate and regulate secondary metabolite synthesis, thus strengthening the plant's innate immune response (Table 1) (Li et al., 2018; Liang et al., 2020; Xia et al., 2022). Furthermore, AP2/ERF TFs have also been implicated in modulating phytohormone signaling, such as jasmonic acid, abscisic acid, auxin, etc., by activating integrated signaling responses to regulate plant growth and defense response (Tweneboah and Oh, 2017; Huang et al., 2016; Feng et al., 2020). Additionally, AP2/ERF TFs coordinate the expression of other TFs such as MYB, WRKY, NAC, and DREB after interacting with specific DNA elements such as ethylene-responsive elements (EREs) and dehydration-responsive elements (DREs) (Wang et al., 2018; Liu et al., 2015).

A large body of literature has corroborated that the interaction between NPR1 and AP2/ERF TFs plays a decisive role in regulating plant growth and defense response under HT stress (Zhang et al., 2012; Zhao et al., 2020; Sun et al., 2022). The interaction between NPR1 and AP2/ERF occurs when the interactive domain of activated NPR1 comes in contact with AP2/ERF (Windram et al., 2014; Wilson et al., 2008; Ma et al., 2022). The interacted NPR1-AP2/ERF TFs stimulate coordinated transcriptional regulation of stress-responsive genes/proteins (Zhang et al., 2017; Yang et al., 2020). The functional mechanism behind the above notion corresponds to the fact that AP2/ERF acts as a downstream regulator of NPR1, thus modulating the plant growth and defense response under HT stress (Yuan et al., 2018; Tu et al., 2020; Li et al., 2023). This coordination between NPR1-AP2/ERF also significantly improves plant immune response under pathogen attack by regulating the expression of PR genes (Wilson et al., 2008; Sun et al., 2020; Liu et al., 2015). The PR genes are responsible for activating the biosynthesis of antimicrobial proteins, enzymes, and secondary metabolites, thus strengthening the immune system of plants under pathogenic attack (Yang et al., 2013; Liu et al., 1998).

Unlike SA, NO does not directly interact with AP2/ERF; works of literature evident that NO and AP2/ERF coordinate via mediating cross-talk with various phytohormones and other signaling pathways (Sun et al., 2018; Zhao et al., 2020). As previously discussed, NO is synthesized in plants via enzymatic and non-enzymatic pathways, and once generated, it can modulate post-translational modifications (PTMs) (Wang et al., 2019; Gupta et al., 2022; Zhang et al., 2023). These NO-mediated PTMs, such as tyrosine nitration and S-nitrosylation, stimulate the binding of NO with specific protein residues (Siddiqui et al., 2017; Astier et al., 2012; Yang et al., 2023). After interacting, NO modulates the functions of various proteins, including AP2/ERF TFs, by regulating their binding with specific cis-regulatory elements (Yin et al., 2013; Yang et al., 2020; Tripathi et al., 2014). Studies have contemplated that NO regulates the activity of AP2/ERF TFs by activating downstream signaling pathways (Srivastava et al., 2019; Wilson et al., 2008; Chakraborty and Acharya, 2017). For instance, NO-mediated regulation of ethylene signaling has been shown to activate AP2/ERF TF genes, thus improving the growth and development of Arabidopsis plants (Yin et al., 2013; Qiao and Fan, 2008; Tripathi et al., 2014). Recent findings have also suggested that NO stimulates the function of AP2/ERF TFs by interacting and regulating JA and ABA signaling, thus improving HT stress response in plants (Feng et al., 2020; Zhang et al., 2023).

**Table 1** SA and NO-mediated transcriptional regulation of heat stress tolerance in plants

| SA and NO mediated Transcriptional Regulation of Heat Stress | Gene/Transcription Factor | Function | SA-Mediated Regulation | NO-Mediated Regulation | References |
|---|---|---|---|---|---|
| **Heat Shock Transcription Factors (HSFs)** | HSFA1 | Regulates the expression of heat shock proteins (HSPs) and other heat stress-responsive genes | SA induces the expression of HSFA1, leading to the upregulation of HSPs and enhanced heat stress tolerance | NO enhances the DNA-binding activity of HSFA1 and promotes the transcriptional activation of HSP genes | Mishra et al. (2018) |
| | HSFA2 | Induces the expression of small heat shock proteins (sHSPs) | SA upregulates the expression of HSFA2, resulting in increased levels of sHSPs and improved heat stress tolerance | NO activates HSFA2 through the MAPK signaling pathway, leading to the induction of sHSPs | Cho et al. (2013) |
| | HSFA3 | Involved in the regulation of acquired thermotolerance | SA enhances the expression of HSFA3, promoting the development of acquired thermotolerance | NO increases the nuclear localization of HSFA3 and augments acquired thermotolerance | Kumar et al. (2009) |

| | | | | | |
|---|---|---|---|---|---|
| | HSFs | Coordinate the expression of multiple HSP genes | SA activates various HSFs, leading to the upregulation of multiple HSP genes | NO regulates the expression of HSFs, resulting in the coordinated induction of HSP genes | Lai et al. (2018) |
| | | | SA and NO interact to modulate the expression and activity of HSFs, synergistically promoting the transcriptional regulation of HSP genes | | |
| **Dehydration-Responsive Element-Binding (DREB) Transcription Factors** | DREB2A | Regulates the expression of stress-responsive genes, including those involved in heat stress tolerance | SA induces the expression of DREB2A, resulting in the upregulation of stress-responsive genes, enhancing heat stress tolerance | NO stabilizes DREB2A protein, prolonging its activity and promoting the transcriptional activation of stress-responsive genes | Xu et al. (2010) |
| | DREB1A/CBF3 | Involved in the regulation of early response to heat stress | SA upregulates the expression of DREB1A/CBF3, leading to improved early heat stress response | NO promotes the nuclear accumulation of DREB1A/CBF3 and enhances the early response to heat stress | Palavan-Unsal and Arisan, (2009) |

| | | | | | |
|---|---|---|---|---|---|
| | | | SA and NO cooperatively regulate the expression of DREB transcription factors, enhancing the transcriptional regulation of stress-responsive genes under heat stress conditions | | |
| **WRKY Transcription Factors** | WRKY25 | Regulates the expression of genes involved in heat stress response | SA upregulates the expression of WRKY25, enhancing heat stress tolerance | NO enhances the DNA-binding activity of WRKY25 and promotes its transcriptional activation | Palavan-Unsal and Arisan, (2009) |
| | WRKY33 | Participates in the crosstalk between different signaling pathways, including SA and ABA, in response to heat stress | SA induces the expression of WRKY33, regulating the crosstalk between SA and ABA signaling pathways for improved heat stress tolerance | NO modulates the expression of WRKY33, mediating the integration of multiple signaling pathways in response to heat stress | Diao et al., (2020) |
| | | | SA and NO synergistically regulate the expression of WRKY transcription factors, facilitating the integration of multiple signaling pathways for enhanced heat stress tolerance | | |

**Table 2** SA and NO-mediated transcriptional regulation of heat stress tolerance in plants

| Role of SA and NO in Regulating Transcription factors | Transcription Factor | Function | SA-Mediated Regulation | NO-Mediated Regulation | References |
|---|---|---|---|---|---|
| **NAC Family** | ANAC019/ANAC055 | Regulates heat stress-responsive genes and enhances heat stress tolerance | SA induces the expression of ANAC019/ANAC055, promoting the upregulation of heat stress-responsive genes and improving heat stress tolerance | NO promotes the nuclear localization of ANAC019/ANAC055 and enhances their transcriptional activity, resulting in the activation of heat stress-responsive genes | Jiang et al. (2020) |
| | ANAC072 | Activates the expression of stress-responsive genes and enhances heat stress tolerance | SA upregulates the expression of ANAC072, leading to the induction of stress-responsive genes and improved heat stress tolerance | NO interacts with ANAC072 to enhance its DNA-binding activity and transcriptional activation, resulting in the upregulation of stress-responsive genes | Nakashima et al. (2006, 2007) |

| | | | | | |
|---|---|---|---|---|---|
| **MYB Family** | AtMYB2 | Regulates the expression of heat shock proteins (HSPs) and improves heat stress tolerance | SA induces the expression of AtMYB2, resulting in the upregulation of HSPs and enhanced heat stress tolerance | NO enhances the binding affinity of AtMYB2 to its target promoters and promotes its transcriptional activation, leading to the induction of HSP genes | Singh et al. (2014) |
| | MYB60 | Participates in the crosstalk between different signaling pathways in response to heat stress | SA regulates the expression of MYB60 and its interaction with other transcription factors, facilitating the integration of multiple signaling pathways | NO modulates the phosphorylation status of MYB60 and enhances its protein stability, enabling the coordination of multiple signaling pathways in response to heat stress | Ding et al. (2008) |
| **MYC Family** | AtMYC2 | Regulates the expression of genes involved in stress responses, including heat stress tolerance | SA enhances the stability of AtMYC2 and its interaction with other co-regulators, leading to the activation of stress-responsive genes | NO interacts with AtMYC2 and promotes its protein accumulation and transcriptional activity, resulting in the upregulation of stress-responsive genes | Zhao et al. (2004) |

| | | | | | |
|---|---|---|---|---|---|
| | MYC3/MYC4 | Participates in the crosstalk between different signaling pathways, including SA and ABA, in response to heat stress | SA upregulates the expression of MYC3/MYC4 and their interaction with other transcription factors, facilitating the integration of multiple signaling pathways | NO regulates the stability and activity of MYC3/MYC4, enabling the integration of multiple signaling pathways and the modulation of heat stress responses | Seo et al. (2009) |
| | | | SA and NO synergistically regulate the expression and activity of MYB transcription factors, enabling the integration of multiple signaling pathways | | |

## DREB

Dehydration Responsive Element Binding (DREB) is a TF protein belonging to the group of AP2/ERF TF family (Liu et al., 1998). Correspondingly, DREB TFs are also involved in the regulation of plant response to various biotic/abiotic stresses (Mizoi et al., 2012; Nakashima et al., 2014). DREB TFs are activated upon the exposure of plants to stress conditions, which stimulate the binding of DREB TFs to cis-regulatory elements, i.e., dehydration-responsive element (DRE)/C-repeat (CRT), in the promoter region of stress-responsive genes (Yamaguchi-Shinozaki and Shinozaki, 2005; Lata and Prasad, 2011). In addition, the activated DREB TFs modulate the transcription of various downstream stress signaling genes involved in growth and adaptation in plants (Mizoi et al., 2012; Qin et al., 2020). These downstream signaling genes often encode proteins such as osmoprotectants, Late Embryogenesis Abundant (LEA) proteins, and other antimicrobial proteins involved in stimulating defense response in plants (Huang et al., 2012; Romero-Puertas et al., 2013).

SA signaling has been known to provoke the activation of several groups of transcription factors, including DREB TFs, by previously known mechanisms (Vlot et al., 2009; Zhang et al., 2012). The activated monomeric NPR1 interacts with DREB TFs in the nucleus, thus enhancing the transcriptional activity of DREB TFs (Schippers et al., 2008; Huang et al., 2016). Concomitantly, the N-terminal of NPR1 plays a mechanistic role in mediating interaction with DREB TF, thus forming the NPR1-DREB complex. The complex then interacts and binds a specific DNA sequence known as DRE or GCC box with the conserved motif of ACCGAC within the promoter region of stress-responsive genes (Liu et al., 1998). The binding of the NPR1-DREB complex to GCC boxes allows the complex to act as a co-activator of other regulatory proteins to mediate the transcriptional activation of different downstream signaling genes (Phukan et al., 2017). Examples of NPR1-DREB complex-mediated transcriptional activation of genes include membrane damage protection, osmoprotectants, antioxidant defense, and other stress-defense proteins (Wang et al., 2018; Pandey and Gautam, 2020).

Increasing literature has suggested that the interaction between NO and DREB TFs plays a crucial role in improving growth and physiology as well as the adaptability of crop plants under HT stress (Feng et al., 2019; Romero-Puertas et al., 2013). NO generated either by enzymatic or non-enzymatic pathways has been shown to modulate the activity of DREB TFs. However, the exact mechanism of NO activating the DREB TFs is still far-fetched. Still, the plant scientist believed the NO-mediated activation of DREB is mediated through protein-protein interaction and associated PTMs (Tiwari et al., 2021; Ding et al., 2019). Studies have contemplated that NO can also influence protein-protein interaction by acting as cofactors in many signaling pathways, including TFs. This NO-mediated protein interaction may also activate DREB TFs, thus ameliorating HT stress-induced oxidative damage in plants (Tu et al., 2020). NO-mediated PTMs such as tyrosine nitration (attachment of $NO_2$ to tyrosine) and S-nitrosylation (attachment of NO to cysteine residues) are also known to affect the structure and function of DREB TFs in plants under stress conditions (Cao et al., 2019; He et al., 2012). Comprehensively, NO-mediated protein-protein interaction, PTMs and interaction with other phytohormones can explicitly regulate the functional mechanism of DREB TFs in plants under various stress conditions.

## NAC

NAC TFs are one of the most prominent families of plant TFs which regulate various physiological and biochemical processes in plants under stress conditions (Puranik et al., 2012; Xu et al., 2014). NAC TFs explicitly modulate the expression of defense-responsive genes allowing plants to adapt and thrive under harsh environments (Rai et al., 2020). NAC TFs function by binding to NAC recognition sites in the promoter region of target genes, thus activating the expression of stress-responsive genes (Olsen et al., 2005; Puranik et al., 2012). Various NAC TFs have been identified in plants that may act as positive/negative modulators of plant responses. For instance, the NAC genes SNAC1 and RD26 have been implicated in playing a role in drought stress tolerance in plants, whereas ANAC019 and ANAC055 regulate salt stress (Nakashima et al., 2007; Fujita et al., 2011; Tran et al., 2020). Further, NAC TFs have also been known to exert pathogen defense response in plants by stimulating the biosynthesis of antimicrobial proteins (Puranik et al., 2012). NAC TFs per se, ANAC055 and ANAC072 are activated in plants in response to pathogen attack and regulate the biosynthesis of PR proteins (Chen et al., 2013; Xu et al., 2014).

The modulation of NAC TFs by SA and NPR1 plays a unique role in activating plant defense response under HT stress conditions (He et al., 2012; Meng et al., 2022). As plants perceive stress signals, the level of SA increases dynamically, which then acts as a signaling molecule under the influence of NPR1 (Zhao et al., 2020; Serba et al., 2021). Exaggerated SA level leads to the activation of NPR1 either by inducing redox changes in NPR1 or by mediating protein-protein interaction within the nucleus (Zhao et al., 2020; Serba et al., 2021). In the middle, the activated NPR1 interacts with NAC transcription factors. The interaction between NPR1 and NAC TFs is mediated by physical protein-protein interactions (He et al., 2012; Serba et al., 2021). The specific domains and regions involved in this interaction have not been fully characterized. Still, it is known that NPR1 interacts with the DNA-binding domain or other areas of NAC TFs (He et al., 2012; Zhou et al., 2000). The interaction between NPR1 and NAC TFs leads to the activation of NAC-dependent gene expression. NAC TFs bind to specific DNA sequences in the promoter regions of target genes (He et al., 2012; Meng et al., 2022). These particular DNA sequences are known as NAC binding elements (NACBs) or NAC recognition sites. The binding of NAC TFs to these elements results in the transcriptional activation of downstream genes (He et al., 2012; Meng et al., 2022).

NO is a gaseous signaling molecule that plays a vital role in diverse physiological processes in plants, including stress responses. Under stress conditions, NO can react with RNS and ROS in the presence of peroxynitrite ($ONOO^-$) to form highly reactive intermediates (Mur et al., 2013). These intermediates can lead to post-translational modifications, including tyrosine nitration (Mata-Pérez et al., 2016). Tyrosine nitration involves adding a nitro group ($^-NO_2$) to tyrosine residues in proteins. Tyrosine nitration can activate NAC transcription factors by inducing conformational changes in the NAC proteins, altering their activity and DNA-binding properties (Hu et al., 2006). The activated NAC transcription factors, under the influence of tyrosine nitration, bind to specific DNA sequences in the promoter regions of stress-responsive genes (Zhou et al., 2000). These DNA sequences, known as NAC binding elements (NACBs) or NAC recognition sites, are typically found in the regulatory regions of stress-responsive genes (Zhou et al., 2023). Once bound to the DNA, the activated NAC transcription factors promote the transcriptional activation of target genes (Jian et al., 2021). These genes are involved in various stress-related processes, including osmotic regulation, antioxidant defense, hormone signaling, and stress tolerance (Jin et al., 2022; Ma et al., 2022).

## WRKY

WRKY TFs (WRKY transcription factors) are a class of plant proteins that play a crucial role in regulating gene expression. TFs are proteins that bind to specific DNA sequences, known as cis-regulatory elements, in the promoter regions of genes, activating or repressing their transcription. They are named after the conserved amino acid sequence WRKYGQK, which is a characteristic feature of this family of TFs (Rushton et al., 2010). They are unique to plants and are classified as members of the more prominent TF family called zinc finger proteins. The zinc finger motif in WRKY TFs enables them to bind to specific DNA sequences in a sequence-specific manner (Rushton et al., 2010). One of the notable features of WRKY TFs is their involvement in plant defense responses. They are critical players in the signaling pathways that mediate plant defense against various biotic stresses, such as pathogen attacks and insect herbivory. When a plant is exposed to such stresses, WRKY TFs are activated and bind to specific DNA sequences in the promoter regions of defense-related genes, leading to their upregulation and subsequent activation of defense mechanisms (Schluttenhofer et al., 2014; Tripathi et al., 2014).

Under high-temperature (HT) stress, plants activate signaling pathways involving salicylic acid (SA) and non-expression of pathogenesis-related genes 1 (NPR1), which ultimately lead to the activation of WRKY signaling. This cascade of molecular events is critical for plants to withstand the detrimental effects of HT stress (Fang et al., 2018). When plants experience HT stress, they produce increased levels of SA, an important signaling molecule involved in various physiological processes, including defense responses against biotic and abiotic stresses (Rivas-San Vicente and Plasencia, 2011). SA acts as a critical regulator in activating the expression of heat stress-responsive genes. It initiates SA signaling by binding to and modifying the activity of specific receptor proteins (Fu et al., 2012). One of the critical components in SA signaling is NPR1, which functions as a transcriptional co-activator. Under normal conditions, NPR1 is present in the cytoplasm in an inactive form. However, in response to HT stress and SA accumulation, NPR1 undergoes a series of modifications that promote its translocation into the nucleus. This translocation is mediated by changes in the redox status of NPR1 and its interaction with specific proteins (Zhang et al., 2012). Once NPR1 reaches the nucleus, it interacts with other transcription factors and co-activators, including members of the WRKY TF family. This interaction leads to the activation of WRKY signaling (Jiang et al., 2016). The precise mechanisms by which NPR1 activates WRKY TFs are still under investigation, but it is believed that NPR1 acts as a bridge, connecting SA signaling to the downstream WRKY TFs (Zhou et al., 2000).

Another critical player in this process is the mitogen-activated protein kinase (MAPK) signaling pathway. MAPKs are activated in response to HT stress and SA signaling, and they play a crucial role in relaying the stress signal to downstream components (Tena et al., 2001). Phosphorylated MAPKs can directly phosphorylate WRKY TFs or other regulatory proteins that, in turn, activate the WRKY TFs (Phukan et al., 2016). This phosphorylation step enhances the DNA-binding affinity and transcriptional activity of WRKY TFs. Activated WRKY TFs bind to specific DNA sequences, known as W-box elements, in the promoter regions of target genes (Rushton et al., 2010). These target genes involve stress-responsive pathways, such as heat shock proteins, antioxidants, and other protective molecules. By binding to the promoter regions of these genes, WRKY TFs enhance their transcriptional activity, leading to increased production of protective proteins and molecules that help plants tolerate HT stress (Feng et al., 2020). The activation of WRKY signaling under HT stress is a complex and coordinated process involving SA, NPR1, and MAPK signaling pathways. SA accumulation triggers events that lead to the nuclear translocation of NPR1, which then interacts with WRKY TFs (Tripathi et al., 2014). Concurrently, MAPKs are activated and phosphorylate WRKY TFs or their regulators, enhancing WRKY TF activity. This culminates in binding activated WRKY TFs to the promoter regions of target genes, promoting the expression of stress-responsive genes (Liu et al., 2015).

Upon HT stress exposure, plants activate signaling pathways involving NO, activating WRKY signaling. The activation of WRKY signaling under HT stress is crucial for plants to cope with the adverse effects of high temperatures. This process involves a series of molecular events that regulate gene expression and contribute to plant stress responses. When exposed to HT stress, plants produce increased levels of NO, which serves as a signaling molecule (Kolbert et al., 2019). NO is involved in various physiological processes, including defense responses against biotic and abiotic stresses. In the context of HT stress, NO acts as a critical regulator in activating the expression of heat stress-responsive genes (Rai et al., 2020). Once NO is produced, it initiates a signaling cascade that leads to the activation of WRKY signaling. One of the critical components involved in this process is S-nitrosylation, which refers to the reversible addition of a NO group to the thiol (-SH) group of specific cysteine residues in target proteins (Lamattina et al., 2003). S-nitrosylation can modulate the activity of target proteins and regulate their function. Under HT stress, NO can S-nitrosylate and modify the activity of specific proteins, including transcription factors and signaling components (Astier et al., 2012). This includes members of the WRKY TF family. S-nitrosylation of WRKY TFs can enhance their DNA-binding affinity and transcriptional activity, promoting the expression of stress-responsive genes (Astier et al., 2012).

In addition to S-nitrosylation, NO can regulate WRKY signaling through interactions with other molecules and signaling pathways. For example, NO can modulate the activity of MAPKs, which are involved in various stress signaling pathways (Wang et al., 2018). MAPKs can be activated by NO through a cascade of phosphorylation events, and subsequently, phosphorylated MAPKs can directly or indirectly activate WRKY TFs (Salazar et al., 2015). Furthermore, NO can interact with other signaling molecules, such as SA and ROS, to coordinate stress responses. NO can enhance SA signaling by promoting the accumulation of SA and activating SA-related pathways (Huang et al., 2023). It can also interact with ROS, forming peroxynitrite, a highly reactive molecule that can modulate gene expression and signaling events involved in stress responses (Rather et al., 2020).

## HSF

Heat Shock Factors (HSFs) play a crucial role in the stress response of plants, particularly in the context of heat stress. HSFs are a family of transcription factors responsible for coordinating the expression of heat shock proteins (HSPs) and other stress-responsive genes (Scharf et al., 2012). They are essential for plant survival and adaptation to high-temperature conditions. When plants are exposed to heat stress, HSFs become activated and undergo various molecular changes to initiate the transcriptional response (Ohama et al., 2016). Activation of HSFs is primarily triggered by the accumulation of misfolded or denatured proteins due to the heat-induced disruption of cellular homeostasis (Fragkostefanakis et al., 2015). The precise mechanisms by which HSFs are activated involve a series of complex molecular events. Under normal conditions, HSFs are present in an inactive state, typically as monomers (Nover et al., 2001). Upon exposure to heat stress, these monomeric HSFs undergo trimerization, where three monomers associate and form a functional HSF trimer (Scharf et al., 2012). Trimerization is a critical step for activating HSFs, as it enables them to bind to specific DNA sequences, known as heat shock elements (HSEs), located in the promoter regions of target genes (Nover et al., 2001).

Once bound to HSEs, the activated HSF trimers act as transcription factors and recruit additional co-regulators and components of the transcriptional machinery (Guo et al., 2016). They facilitate the assembly of a transcriptional complex that promotes the expression of heat shock genes, including various classes of HSPs (Nover et al., 2001). These HSPs function as molecular chaperones, helping to stabilize and refold damaged proteins or targeting them for degradation to restore cellular proteostasis. Apart from their role in activating HSP genes, HSFs are also involved in regulating other stress-responsive genes. They can interact with and modulate the expression of genes involved in antioxidant defenses, signaling pathways, and metabolic adjustments (Scharf et al., 2012). For example, HSFs can interact with components of the ROS signaling pathway and contribute to the regulation of ROS scavenging enzymes. HSFs can also regulate the expression of genes involved in the biosynthesis and signaling of plant hormones, such as ABA and SA. By coordinating the expression of a wide range of stress-responsive genes, HSFs contribute to plants' overall stress tolerance and survival under adverse conditions (Fragkostefanakis et al., 2015). Their role in regulating gene expression networks involved in stress responses is essential for plant adaptation and resilience to environmental challenges.

When exposed to HT stress, plants produce increased levels of SA, a key signaling molecule involved in various physiological processes, including defense responses against biotic and abiotic stresses (Vlot et al., 2009). SA is a crucial regulator in activating the expression of heat stress-responsive genes. It initiates SA signaling by binding to and modifying the activity of specific receptor proteins (Fu et al., 2012). Once in the nucleus, NPR1 interacts with other transcription factors and co-activators, including HSFs. This interaction leads to the activation of HSF signaling (Pajerowska-Mukhtar et al., 2012). The precise molecular mechanisms by which NPR1 activates HSFs are still under investigation, but it is believed that NPR1 acts as a bridge, connecting SA signaling to the downstream HSFs (Després et al., 2003). The interplay between SA, NPR1, and HSFs is further mediated by MAPK signaling pathways. MAPKs are involved in various stress signaling pathways, and SA and other stress-related signals can trigger their activation (Wang et al., 2004). MAPKs can phosphorylate and activate NPR1, which, in turn, interacts with HSFs to activate their transcriptional activity (Zhang et al., 2007). Once activated, HSFs form trimers and bind to specific DNA sequences known as HSEs located in the promoter regions of target genes (Mishra et al., 2002). This binding facilitates the recruitment of co-regulators and components of the transcriptional machinery, leading to the expression of heat shock genes, including various classes of HSPs (Mittler et al., 2012). HSPs function as molecular chaperones, aiding in protein folding, stabilization, and degradation, thus restoring cellular proteostasis under HT stress conditions (Wang et al., 2004).

NO production under HT stress occurs through the activity of enzymes called nitric oxide synthases (NOS) (Simontacchi et al., 2015). These enzymes catalyze the conversion of L-arginine to NO and citrulline. The increase in NO levels serves as an early signal indicating the presence of HT stress in plants (Mittler et al., 2012). Once NO is produced, it initiates a signaling cascade that leads to the activation of HSF signaling. One of the key molecular events involved in NO-mediated HSF activation is the S-nitrosylation of HSFs. S-nitrosylation refers to the reversible addition of a NO group to the thiol (-SH) group of specific cysteine residues in target proteins (Tousoulis et al., 2012). S-nitrosylation can modulate the activity of target proteins and regulate their function. Under HT stress, NO can S-nitrosylate and modify the activity of specific proteins, including HSFs (Feng et al., 2019). S-nitrosylation of HSFs can enhance their DNA-binding affinity and transcriptional activity, promoting the expression of heat stress-responsive genes. In addition to S-nitrosylation, NO can regulate HSF signaling through interactions with other molecules and signaling pathways. NO can modulate the activity of MAPKs, which are involved in various stress signaling pathways (Xuan et al., 2012). Phosphorylated MAPKs can directly or indirectly activate HSFs, enhancing their transcriptional activity and contributing to the expression of heat stress-responsive genes (Wang et al., 2004).

## MYC/MYB

MYB (Myeloblastosis) and MYC (Myelocytomatosis) are two transcription factors that play important roles in regulating various processes in plants (Dubos et al., 2010). These transcription factors are involved in diverse biological processes, including growth and development, hormone signaling, response to environmental stresses, and secondary metabolism (Ambawat et al., 2013; Cao et al., 2016). The MYB family of transcription factors is characterized by a highly conserved MYB DNA-binding domain consisting of one to four repeats of approximately 50 amino acids (Du et al., 2012). Each repeat contains three alpha helices that form a helix-turn-helix structure, allowing MYB proteins to bind to specific DNA sequences (Ogata et al., 1994). Additionally, MYB transcription factors significantly regulate secondary metabolites, including flavonoids, anthocyanins, and lignin (Dubos et al., 2010). They control the expression of genes involved in the biosynthesis of these compounds, which contribute to pigmentation, UV protection, and defense against pathogens and herbivores.

On the other hand, MYC transcription factors are members of the basic helix-loop-helix (bHLH) family of transcription factors. They function as critical regulators of cell proliferation, differentiation, and responses to various environmental cues in plants. MYC proteins form heterodimers with bHLH partners, such as members of the bHLH family transcription factors called bHLH (E) proteins, to bind specific DNA sequences known as E-box elements. These interactions enable MYC proteins to regulate the expression of target genes involved in cell cycle control, hormone signaling, stress responses, and metabolism (Paponov et al., 2008; Sornaraj et al., 2016; Cao et al., 2019). In plant growth and development, MYC transcription factors regulate processes such as leaf development, root growth, shoot branching, and flowering time (Lamattina et al., 2003). They also control hormone signaling pathways, particularly in response to abiotic and biotic stresses (Lamattina et al., 2003).

When exposed to HT stress, plants produce increased levels of SA, a key signaling molecule involved in various physiological processes, including defense responses against biotic and abiotic stresses (Rivas-San Vicente and Plasencia, 2011; Miura and Tada, 2014). SA is a crucial regulator in activating MYB/MYC signaling and modulating gene expression under HT stress conditions (Astier et al., 2012). The activation of MYB/MYC signaling under HT stress also involves regulating gene expression and modulating specific molecular pathways (Tu et al., 2020; Wilson et al., 2008). SA and NPR1 signaling can influence the expression of MYB/MYC genes directly or indirectly (Zhang et al., 2012; Wang et al., 2022). These transcription factors can then bind to specific DNA sequences in the promoter regions of target genes, regulating their expression (Mengiste et al., 2003; Dubos et al., 2010). MYB transcription factors are known to be involved in various stress-responsive pathways, including heat stress. They can handle gene expression in heat shock responses, osmotic regulation, and antioxidant defense (Huang et al., 2016). By activating MYB signaling, SA and NPR1 signaling can modulate the expression of these stress-responsive genes, allowing plants to adapt and tolerate HT stress.

Additionally, SA and NPR1 signaling can interact with other signaling pathways, such as MAPK cascades, which are known to regulate

MYB/MYC signaling (Liu and Zhang, 2020; Rai et al., 2020). MAPKs can be activated by SA and other stress-related signals (Wang et al., 2018; Liu et al., 2015). They can phosphorylate and activate MYB/MYC transcription factors, enhancing their transcriptional activity and contributing to the expression of stress-responsive genes (Yin et al., 2013; Meng et al., 2022).

NO production under HT stress occurs through the activity of enzymes called NOS (Rai et al., 2020). These enzymes catalyze the conversion of L-arginine to NO and citrulline. The increase in NO levels serves as an early signal indicating the presence of HT stress in plants (Gupta et al., 2022). Once NO is produced, it initiates a signaling cascade that activates MYB/MYC signaling (Qiao and Fan 2008). One of the key molecular events involved in NO-mediated MYB/MYC activation is the S-nitrosylation of target proteins (Corpas et al., 2011). S-nitrosylation refers to the reversible addition of a NO group to the thiol (-SH) group of specific cysteine residues in target proteins. S-nitrosylation can modulate the activity of target proteins and regulate their function (Lindermayr and Durner, 2009). Under HT stress, NO can S-nitrosylate and modify the activity of specific proteins, including transcription factors such as MYB and MYC (Huang et al., 2022). S-nitrosylation of MYB/MYC transcription factors can enhance their DNA-binding affinity and transcriptional activity, promoting the expression of stress-responsive genes (Qiao and Fan, 2008). In addition to S-nitrosylation, NO can regulate MYB/MYC signaling through interactions with other molecules and signaling pathways. NO can interact with reactive oxygen species (ROS) to form reactive nitrogen species (RNS), which can further modulate gene expression and signaling pathways (Wei et al., 2017). This interplay between NO and ROS/RNS can activate MYB/MYC signaling and contribute to stress responses under HT conditions (Zhao, 2007).

Furthermore, NO can cross-talk with other signaling molecules, such as phytohormones like ABA, to regulate MYB/MYC signaling. ABA is known to be involved in plant responses to various stresses, including HT stress (Chen et al., 2016). NO and ABA signaling pathways can interact and synergistically activate MYB/MYC transcription factors, leading to the expression of stress-responsive genes (Nazar et al., 2017).

## Conclusion

Plants are continuously exposed to various stresses, including high-temperature (HT) stress, which poses significant challenges to their growth and survival. To combat the detrimental effects of HT stress, plants activate complex signaling pathways that regulate gene expression and enable adaptive responses. This abstract provides a comprehensive overview of three key signaling pathways involved in plant responses to HT stress: SA and non-expressor of NPR1 signaling, NO signaling, and HSF signaling. The SA and NPR1 signaling pathway is activated under HT stress conditions and is crucial in plant defence mechanisms. Increased levels of SA trigger the translocation of NPR1 into the nucleus, where it interacts with transcription factors, including MYB and MYC, leading to the activation of stress-responsive genes. This pathway is a central regulator of HT stress responses, coordinating gene expression and enhancing stress tolerance. Similarly, NO signaling has become vital in plant responses to HT stress. NO is produced through the action of NOS and functions as a signaling molecule. It activates MYB/MYC transcription factors through S-nitrosylation, enhancing their DNA-binding affinity and promoting the expression of stress-responsive genes. NO signaling also interacts with ROS, phytohormones like ABA, and other signaling pathways to amplify stress responses and enhance thermotolerance.

The HSF signaling pathway is also critical for plant adaptation to HT stress. HSFs act as transcription factors that regulate the expression of HSPs, which are crucial for protein homeostasis under HT conditions. Understanding the molecular mechanisms underlying these signaling pathways is essential for deciphering plant responses to HT stress and developing strategies to improve crop stress tolerance. Further research will unravel the intricate interplay between these signaling pathways, identify vital regulators and downstream targets, and pave the way for innovative approaches to enhance crop resilience to HT stress, ensuring sustainable agricultural productivity in climate change.

## References

Alamri, S.A., Siddiqui, M.H., Al-Khaishany, M.Y., Khan, M.N., Ali, H.M. and Alakeel, K.A., 2019. Nitric oxide-mediated cross-talk of proline and heat shock proteins induce thermotolerance in *Vicia faba* L. Environmental and Experimental Botany, 161, pp.290-302.

Ambawat, S., Sharma, P., Yadav, N.R. and Yadav, R.C., 2013. MYB transcription factor genes as regulators for plant responses: an overview. Physiology and Molecular Biology of Plants, 19, pp.307-321.

Asgher, M., Per, T.S., Masood, A., Fatma, M., Freschi, L., Corpas, F.J. and Khan, N.A., 2017. Nitric oxide signaling and its crosstalk with other plant growth regulators in plant responses to abiotic stress. Environmental Science and Pollution Research, 24, pp.2273-2285.

Astier, J., Kulik, A., Koen, E., Besson-Bard, A., Bourque, S., Jeandroz, S., Lamotte, O. and Wendehenne, D., 2012. Protein S-nitrosylation: what's going on in plants?. Free Radical Biology and Medicine, 53(5), pp.1101-1110.

Bellin, D., Asai, S., Delledonne, M. and Yoshioka, H., 2013. Nitric oxide as a mediator for defense responses. Molecular plant-microbe interactions, 26(3), pp.271-277.

Bhargava, S. and Sawant, K., 2013. Drought stress adaptation: metabolic adjustment and regulation of gene expression. Plant breeding, 132(1), pp.21-32.

Cao, N., Zhan, B. and Zhou, X., 2019. Nitric oxide as a downstream signaling molecule in brassinosteroid-mediated virus susceptibility to maize chlorotic mottle virus in maize. Viruses, 11(4), p.368.

Cao, Y., Han, Y., Li, D., Lin, Y. and Cai, Y., 2016. MYB transcription factors in Chinese pear (Pyrus bretschneider i Rehd.): Genome-wide identification, classification, and expression profiling during fruit development. Frontiers in plant science, 7, p.577.

Chakraborty, N. and Acharya, K., 2017. “NO way”! Says the plant to abiotic stress. Plant Gene, 11, pp.99-105.

Chen, G., Greer, M.S. and Weselake, R.J., 2013. Plant phospholipase A: advances in molecular biology, biochemistry, and cellular function. BioMolecular Concepts, 4(5), pp.527-532.

Chen, J., Yu, F., Liu, Y., Du, C., Li, X., Zhu, S., Wang, X., Lan, W., Rodriguez, P.L., Liu, X. and Li, D., 2016. FERONIA interacts with ABI2-type phosphatases to facilitate signaling cross-talk between abscisic acid and RALF peptide in Arabidopsis. Proceedings of the National Academy of Sciences, 113(37), pp.E5519-E5527.

Chen, M., Gan, L., Zhang, J., Shen, Y., Qian, J., Han, M., Zhang, C., Fan, J., Sun, S. and Yan, X., 2021. A regulatory network of heat shock modules-photosynthesis-redox systems in response to cold stress across a latitudinal gradient in bermudagrass. Frontiers in plant science, 12, p.751901.

Cho, S.M., Kim, Y.H., Anderson, A.J. and Kim, Y.C., 2013. Nitric oxide and hydrogen peroxide production are involved in systemic drought tolerance induced by 2R, 3R-butanediol in *Arabidopsis thaliana*. The plant pathology journal, 29(4), p.427.

Corpas, F.J., Leterrier, M., Valderrama, R., Airaki, M., Chaki, M., Palma, J.M. and Barroso, J.B., 2011. Nitric oxide imbalance provokes a nitrosative response in plants under abiotic stress. Plant Science, 181(5), pp.604-611.

Després, C., Chubak, C., Rochon, A., Clark, R., Bethune, T., Desveaux, D. and Fobert, P.R., 2003. The Arabidopsis NPR1 disease resistance protein is a novel cofactor that confers redox regulation of DNA binding activity to the basic domain/leucine zipper transcription factor TGA1. The Plant Cell, 15(9), pp.2181-2191.

Diao, P., Chen, C., Zhang, Y., Meng, Q., Lv, W. and Ma, N., 2020. The role of NAC transcription factor in plant cold response. Plant Signaling & Behavior, 15(9), p.1785668.

Ding, X., Cao, Y., Huang, L., Zhao, J., Xu, C., Li, X. and Wang, S., 2008. Activation of the indole-3-acetic acid–amido synthetase GH3-8 suppresses expansin expression and promotes salicylate-and jasmonate-independent basal immunity in rice. The Plant Cell, 20(1), pp.228-240.

Ding, Y., Fan, B., Zhu, C. and Chen, Z., 2023. Shared and related molecular targets and actions of salicylic acid in plants and humans. Cells, 12(2), p.219.

Ding, Y., Shi, Y. and Yang, S., 2019. Advances and challenges in uncovering cold tolerance regulatory mechanisms in plants. New Phytologist, 222(4), pp.1690-1704.

Du, H., Feng, B.R., Yang, S.S., Huang, Y.B. and Tang, Y.X., 2012. The R2R3-MYB transcription factor gene family in maize. PloS one, 7(6), p.e37463.

Dubos, C., Stracke, R., Grotewold, E., Weisshaar, B., Martin, C. and Lepiniec, L., 2010. MYB transcription factors in Arabidopsis. Trends in plant science, 15(10), pp.573-581.

Fang, C., Dou, L., Liu, Y., Yu, J. and Tu, J., 2018. Heat stress-responsive transcriptome analysis in heat susceptible and tolerant rice by high-throughput sequencing. Ecological Genetics and Genomics, 6, pp.33-40.

Feng, J., Chen, L. and Zuo, J., 2019. Protein S-nitrosylation in plants: current progresses and challenges. Journal of Integrative Plant Biology, 61(12), pp.1206-1223.

Feng, K., Hou, X.L., Xing, G.M., Liu, J.X., Duan, A.Q., Xu, Z.S., Li, M.Y., Zhuang, J. and Xiong, A.S., 2020. Advances in AP2/ERF super-family transcription factors in plant. Critical reviews in biotechnology, 40(6), pp.750-776.

Fragkostefanakis, S., Roeth, S., Schleiff, E. and SCHARF, K.D., 2015. Prospects of engineering thermotolerance in crops through modulation of heat stress transcription factor and heat shock protein networks. Plant, cell & environment, 38(9), pp.1881-1895.

Fu, Z.Q. and Dong, X., 2013. Systemic acquired resistance: turning local infection into global defense. Annual review of plant biology, 64, pp.839-863.

Fujita, Y., Fujita, M., Shinozaki, K. and Yamaguchi-Shinozaki, K., 2011. ABA-mediated transcriptional regulation in response to osmotic stress in plants. Journal of plant research, 124, pp.509-525.

Gilroy, E. and Breen, S., 2022. Interplay between phytohormone signaling pathways in plant defence–other than salicylic acid and jasmonic acid. Essays in Biochemistry, 66(5), pp.657-671.

Gupta, K.J., Kaladhar, V.C., Fitzpatrick, T.B., Fernie, A.R., Møller, I.M. and Loake, G.J., 2022. Nitric oxide regulation of plant metabolism. Molecular Plant.

Gupta, K.J., Kaladhar, V.C., Fitzpatrick, T.B., Fernie, A.R., Møller, I.M. and Loake, G.J., 2022. Nitric oxide regulation of plant metabolism. Molecular Plant.

He, H.Y., He, L.F., Gu, M.H. and Li, X.F., 2012. Nitric oxide improves aluminum tolerance by regulating hormonal equilibrium in the root apices of rye and wheat. Plant Science, 183, pp.123-130.

Hu, D., Wei, L. and Liao, W., 2021. Brassinosteroids in plants: Crosstalk with small-molecule compounds. Biomolecules, 11(12), p.1800.

Hu, D.G., Sun, C.H., Ma, Q.J., You, C.X., Cheng, L. and Hao, Y.J., 2016. MdMYB1 regulates anthocyanin and malate accumulation by directly facilitating their transport into vacuoles in apples. Plant Physiology, 170(3), pp.1315-1330.

Hu, H., Dai, M., Yao, J., Xiao, B., Li, X., Zhang, Q. and Xiong, L., 2006. Overexpressing a NAM, ATAF, and CUC (NAC) transcription factor enhances drought resistance and salt tolerance in rice. Proceedings of the National Academy of Sciences, 103(35), pp.12987-12992.

Huang, J., Chen, F., Wu, S., Li, J. and Xu, W., 2016. Cotton GhMYB7 is predominantly expressed in developing fibers and regulates secondary cell wall biosynthesis in transgenic Arabidopsis. Science China Life Sciences, 59, pp.194-205.

Huang, J., Sun, S., Xu, D., Lan, H., Sun, H., Wang, Z., Bao, Y., Wang, J., Tang, H. and Zhang, H., 2012. A TFIIIA-type zinc finger protein confers multiple abiotic stress tolerances in transgenic rice (Oryza sativa L.). Plant molecular biology, 80, pp.337-350.

Iqbal, N., Fatma, M., Khan, N.A. and Umar, S., 2019. Regulatory role of proline in heat stress tolerance: modulation by salicylic acid. In Plant Signaling Molecules (pp. 437-448). Woodhead Publishing.

Jian, W., Zheng, Y., Yu, T., Cao, H., Chen, Y., Cui, Q., Xu, C. and Li, Z., 2021. SlNAC6, A NAC transcription factor, is involved in drought stress response and reproductive process in tomato. Journal of Plant Physiology, 264, p.153483.

Jiang, G., Zeng, J., Li, Z., Song, Y., Yan, H., He, J., Jiang, Y. and Duan, X., 2020. Redox regulation of the NOR transcription factor is involved in the regulation of fruit ripening in tomato. Plant Physiology, 183(2), pp.671-685.

Jiang, Y., Qiu, Y., Hu, Y. and Yu, D., 2016. Heterologous expression of AtWRKY57 confers drought tolerance in *Oryza sativa*. Frontiers in Plant Science, 7, p.145.

Jin, J.F., Wang, Z.Q., He, Q.Y., Wang, J.Y., Li, P.F., Xu, J.M., Zheng, S.J., Fan, W. and Yang, J.L., 2020. Genome-wide identification and expression analysis of the NAC transcription factor family in tomato (*Solanum lycopersicum*) during aluminum stress. BMC genomics, 21, pp.1-14.

Kolbert, Z.S., Barroso, J.B., Brouquisse, R., Corpas, F.J., Gupta, K.J., Lindermayr, C., Loake, G.J., Palma, J.M., Petřivalský, M., Wendehenne, D. and Hancock, J.T., 2019. A forty year journey: The generation and roles of NO in plants. Nitric Oxide, 93, pp.53-70.

Kumar, M., Busch, W., Birke, H., Kemmerling, B., Nürnberger, T. and Schöffl, F., 2009. Heat shock factors HsfB1 and HsfB2b are involved in the regulation of Pdf1. 2 expression and pathogen resistance in Arabidopsis. Molecular plant, 2(1), pp.152-165.

Lai, Y.S., Renna, L., Yarema, J., Ruberti, C., He, S.Y. and Brandizzi, F., 2018. Salicylic acid-independent role of NPR1 is required for protection from proteotoxic stress in the plant endoplasmic reticulum. Proceedings of the National Academy of Sciences, 115(22), pp.E5203-E5212.

Lamattina, L., García-Mata, C., Graziano, M. and Pagnussat, G., 2003. Nitric oxide: the versatility of an extensive signal molecule. Annual review of plant biology, 54(1), pp.109-136.

Lata, C. and Prasad, M., 2011. Role of DREBs in regulation of abiotic stress responses in plants. Journal of experimental botany, 62(14), pp.4731-4748.

Li, B., Fan, R., Yang, Q., Hu, C., Sheng, O., Deng, G., Dong, T., Li, C., Peng, X., Bi, F. and Yi, G., 2020. Genome-wide identification and characterization of the NAC transcription factor family in *Musa acuminata* and expression analysis during fruit ripening. International journal of molecular sciences, 21(2), p.634.

Lindermayr, C. and Durner, J., 2009. S-Nitrosylation in plants: pattern and function. Journal of Proteomics, 73(1), pp.1-9.

Lindermayr, C., Sell, S., Müller, B., Leister, D. and Durner, J., 2010. Redox regulation of the NPR1-TGA1 system of *Arabidopsis thaliana* by nitric oxide. The Plant Cell, 22(8), pp.2894-2907.

Liu, Q., Kasuga, M., Sakuma, Y., Abe, H., Miura, S., Yamaguchi-Shinozaki, K. and Shinozaki, K., 1998. Two transcription factors, DREB1 and DREB2, with an EREBP/AP2 DNA binding domain separate two cellular signal transduction pathways in drought-and low-temperature-responsive gene expression, respectively, in Arabidopsis. The Plant Cell, 10(8), pp.1391-1406.

Liu, Q., Wang, Z., Xu, X., Zhang, H. and Li, C., 2015. Genome-wide analysis of C2H2 zinc-finger family transcription factors and their responses to abiotic stresses in poplar (Populus trichocarpa). PloS one, 10(8), p.e0134753.

Lozano-Juste, J. and León, J., 2011. Nitric oxide regulates DELLA content and PIF expression to promote photomorphogenesis in Arabidopsis. Plant Physiology, 156(3), pp.1410-1423.

Ma, Q., Xia, Z., Cai, Z., Li, L., Cheng, Y., Liu, J. and Nian, H., 2019. GmWRKY16 enhances drought and salt tolerance through an ABA-mediated pathway in Arabidopsis thaliana. Frontiers in plant science, 9, p.1979.

Ma, W., Kang, X., Liu, P., She, K., Zhang, Y., Lin, X., Li, B. and Chen, Z., 2022. The NAC-like transcription factor CsNAC7 positively regulates the caffeine biosynthesis-related gene yhNMT1 in Camellia sinensis. Horticulture research, 9, p.uhab046.

Ma, X., Zhang, Y., Turečková, V., Xue, G.P., Fernie, A.R., Mueller-Roeber, B. and Balazadeh, S., 2018. The NAC transcription factor SlNAP2 regulates leaf senescence and fruit yield in tomato. Plant physiology, 177(3), pp.1286-1302.

Mata-Pérez, C., Begara-Morales, J.C., Chaki, M., Sánchez-Calvo, B., Valderrama, R., Padilla, M.N., Corpas, F.J. and Barroso, J.B., 2016. Protein tyrosine nitration during development and abiotic stress response in plants. Frontiers in Plant Science, 7, p.1699.

McCarthy-Suárez, I., 2017. Role of reactive oxygen species in auxin herbicide phytotoxicity: current information and hormonal implications—are gibberellins, cytokinins, and polyamines involved?. Botany, 95(4), pp.369-385.

Meng, L., Yang, H., Xiang, L., Wang, Y. and Chan, Z., 2022. NAC transcription factor TgNAP promotes tulip petal senescence. Plant Physiology, 190(3), pp.1960-1977.

Mengiste, T., Chen, X., Salmeron, J. and Dietrich, R., 2003. The BOTRYTIS SUSCEPTIBLE1 gene encodes an R2R3MYB transcription factor protein that is required for biotic and abiotic stress responses in Arabidopsis. The Plant Cell, 15(11), pp.2551-2565.

Mishra, D., Shekhar, S., Singh, D., Chakraborty, S. and Chakraborty, N., 2018. Heat shock proteins and abiotic stress tolerance in plants. Regulation of heat shock protein responses, pp.41-69.

Mishra, S.K., Tripp, J., Winkelhaus, S., Tschiersch, B., Theres, K., Nover, L. and Scharf, K.D., 2002. In the complex family of heat stress transcription factors, HsfA1 has a unique role as master regulator of thermotolerance in tomato. Genes & development, 16(12), pp.1555-1567.

Mittler, R., Finka, A. and Goloubinoff, P., 2012. How do plants feel the heat?. Trends in biochemical sciences, 37(3), pp.118-125.

Miura, K. and Tada, Y., 2014. Regulation of water, salinity, and cold stress responses by salicylic acid. Frontiers in plant science, 5, p.4.

Mizoi, J., Shinozaki, K. and Yamaguchi-Shinozaki, K., 2012. AP2/ERF family transcription factors in plant abiotic stress responses. Biochimica et Biophysica Acta (BBA)-Gene Regulatory Mechanisms, 1819(2), pp.86-96.

Mur, L.A., Mandon, J., Persijn, S., Cristescu, S.M., Moshkov, I.E., Novikova, G.V., Hall, M.A., Harren, F.J., Hebelstrup, K.H. and Gupta, K.J., 2013. Nitric oxide in plants: an assessment of the current state of knowledge. AoB plants, 5, p.pls052.

Nakashima, K., Fujita, Y., Katsura, K., Maruyama, K., Narusaka, Y., Seki, M., Shinozaki, K. and Yamaguchi-Shinozaki, K., 2006. Transcriptional regulation of ABI3-and ABA-responsive genes including RD29B and RD29A in seeds, germinating embryos, and seedlings of Arabidopsis. Plant molecular biology, 60, pp.51-68.

Nakashima, K., Tran, L.S.P., Van Nguyen, D., Fujita, M., Maruyama, K., Todaka, D., Ito, Y., Hayashi, N., Shinozaki, K. and Yamaguchi-Shinozaki, K., 2007. Functional analysis of a NAC-type transcription factor OsNAC6 involved in abiotic and biotic stress-responsive gene expression in rice. The Plant Journal, 51(4), pp.617-630.

Nakashima, K., Yamaguchi-Shinozaki, K. and Shinozaki, K., 2014. The transcriptional regulatory network in the drought response and its crosstalk in abiotic stress responses including drought, cold, and heat. Frontiers in plant science, 5, p.170.

Nazar, R., Iqbal, N. and Umar, S., 2017. Heat stress tolerance in plants: action of salicylic acid. Salicylic acid: A multifaceted hormone, pp.145-161.

Nover, L., Bharti, K., Döring, P., Mishra, S.K., Ganguli, A. and Scharf, K.D., 2001. Arabidopsis and the heat stress transcription factor world: how many heat stress transcription factors do we need?. Cell stress & chaperones, 6(3), p.177.

Ogata, K., Morikawa, S., Nakamura, H., Sekikawa, A., Inoue, T., Kanai, H., Sarai, A., Ishii, S. and Nishimura, Y., 1994. Solution structure of a specific DNA complex of the Myb DNA-binding domain with cooperative recognition helices. Cell, 79(4), pp.639-648.

Ohama, N., Kusakabe, K., Mizoi, J., Zhao, H., Kidokoro, S., Koizumi, S., Takahashi, F., Ishida, T., Yanagisawa, S., Shinozaki, K. and Yamaguchi-Shinozaki, K., 2016. The transcriptional cascade in the heat stress response of Arabidopsis is strictly regulated at the level of transcription factor expression. The Plant Cell, 28(1), pp.181-201.

Olsen, A.N., Ernst, H.A., Leggio, L.L. and Skriver, K., 2005. NAC transcription factors: structurally distinct, functionally diverse. Trends in plant science, 10(2), pp.79-87.

Pajerowska-Mukhtar, K.M., Wang, W., Tada, Y., Oka, N., Tucker, C.L., Fonseca, J.P. and Dong, X., 2012. The HSF-like transcription factor TBF1 is a major molecular switch for plant growth-to-defense transition. Current Biology, 22(2), pp.103-112.

Palavan-Unsal, N. and Arisan, D., 2009. Nitric oxide signaling in plants. The Botanical Review, 75, pp.203-229.

Pandey, A.K. and Gautam, A., 2020. Stress responsive gene regulation in relation to hydrogen sulfide in plants under abiotic stress. Physiologia plantarum, 168(2), pp.511-525.

Paponov, I.A., Paponov, M., Teale, W., Menges, M., Chakrabortee, S., Murray, J.A. and Palme, K., 2008. Comprehensive transcriptome analysis of auxin responses in Arabidopsis. Molecular Plant, 1(2), pp.321-337.

Parray, J.A., Yaseen Mir, M., Shameem, N., Parray, J.A., Yaseen Mir, M. and Shameem, N., 2019. Stress management: Sustainable approach towards resilient agriculture. Sustainable agriculture: Biotechniques in plant biology, pp.231-270.

Paul, S. and Roychoudhury, A., 2020. Regulation of physiological aspects in plants by hydrogen sulfide and nitric oxide under challenging environment. Physiologia plantarum, 168(2), pp.374-393.

Phukan, U.J., Jeena, G.S. and Shukla, R.K., 2016. WRKY transcription factors: molecular regulation and stress responses in plants. Frontiers in plant science, 7, p.760.

Phukan, U.J., Jeena, G.S., Tripathi, V. and Shukla, R.K., 2017. Regulation of Apetala2/Ethylene response factors in plants. Frontiers in plant science, 8, p.150.

Pieterse, C.M., Van der Does, D., Zamioudis, C., Leon-Reyes, A. and Van Wees, S.C., 2012. Hormonal modulation of plant immunity. Annual review of cell and developmental biology, 28, pp.489-521.

Puranik, S., Sahu, P.P., Srivastava, P.S. and Prasad, M., 2012. NAC proteins: regulation and role in stress tolerance. Trends in plant science, 17(6), pp.369-381.

Qiao, W. and Fan, L.M., 2008. Nitric oxide signaling in plant responses to abiotic stresses. Journal of Integrative Plant Biology, 50(10), pp.1238-1246.

Qiao, W. and Fan, L.M., 2008. Nitric oxide signaling in plant responses to abiotic stresses. Journal of Integrative Plant Biology, 50(10), pp.1238-1246.

Qin, F., Shinozaki, K. and Yamaguchi-Shinozaki, K., 2011. Achievements and challenges in understanding plant abiotic stress responses and tolerance. Plant and Cell Physiology, 52(9), pp.1569-1582.

Rai, K.K., Pandey, N. and Rai, S.P., 2020. Salicylic acid and nitric oxide signaling in plant heat stress. Physiologia plantarum, 168(2), pp.241-255.

Rather, B.A., Mir, I.R., Sehar, Z., Anjum, N.A., Masood, A. and Khan, N.A., 2020. The outcomes of the functional interplay of nitric oxide and hydrogen sulfide in metal stress tolerance in plants. Plant Physiology and Biochemistry, 155, pp.523-534.

Rivas-San Vicente, M. and Plasencia, J., 2011. Salicylic acid beyond defence: its role in plant growth and development. Journal of experimental botany, 62(10), pp.3321-3338.

Rivas-San Vicente, M. and Plasencia, J., 2011. Salicylic acid beyond defence: its role in plant growth and development. Journal of experimental botany, 62(10), pp.3321-3338.

Romero-Puertas, M.C., Rodríguez-Serrano, M. and Sandalio, L.M., 2013. Protein S-nitrosylation in plants under abiotic stress: an overview. Frontiers in Plant Science, 4, p.373.

Rushton, P.J., Somssich, I.E., Ringler, P. and Shen, Q.J., 2010. WRKY transcription factors. Trends in plant science, 15(5), pp.247-258.

Rushton, P.J., Somssich, I.E., Ringler, P. and Shen, Q.J., 2010. WRKY transcription factors. Trends in plant science, 15(5), pp.247-258.

Salazar, C., Hernández, C. and Pino, M.T., 2015. Plant water stress: Associations between ethylene and abscisic acid response. Chilean journal of agricultural research, 75, pp.71-79.

Scharf, K.D., Berberich, T., Ebersberger, I. and Nover, L., 2012. The plant heat stress transcription factor (Hsf) family: structure, function and evolution. Biochimica et Biophysica Acta (BBA)-Gene Regulatory Mechanisms, 1819(2), pp.104-119.

Schippers, J.H., Nunes-Nesi, A., Apetrei, R., Hille, J., Fernie, A.R. and Dijkwel, P.P., 2008. The Arabidopsis old5 mutation of quinolinate synthase affects NAD biosynthesis and causes early ageing. Molecular aspects of ageing and the onset of leaf senescence, p.121.

Schluttenhofer, C., Pattanaik, S., Patra, B. and Yuan, L., 2014. Analyses of *Catharanthus roseus* and Arabidopsis thaliana WRKY transcription factors reveal involvement in jasmonate signaling. BMC genomics, 15(1), pp.1-20.

Seo, P.J., Xiang, F., Qiao, M., Park, J.Y., Lee, Y.N., Kim, S.G., Lee, Y.H., Park, W.J. and Park, C.M., 2009. The MYB96 transcription factor mediates abscisic acid signaling during drought stress response in Arabidopsis. Plant physiology, 151(1), pp.275-289.

Serba, D.D., Meng, X., Schnable, J., Bashir, E., Michaud, J.P., Vara Prasad, P.V. and Perumal, R., 2021. Comparative Transcriptome analysis reveals genetic mechanisms of sugarcane aphid resistance in grain Sorghum. International Journal of Molecular Sciences, 22(13), p.7129.

Sharma, R., De Vleesschauwer, D., Sharma, M.K. and Ronald, P.C., 2013. Recent advances in dissecting stress-regulatory crosstalk in rice. Molecular Plant, 6(2), pp.250-260.

Simontacchi, M., Galatro, A., Ramos-Artuso, F. and Santa-María, G.E., 2015. Plant survival in a changing environment: the role of nitric oxide in plant responses to abiotic stress. Frontiers in plant science, 6, p.977.

Singh, N., Mishra, A. and Jha, B., 2014. Over-expression of the peroxisomal ascorbate peroxidase (SbpAPX) gene cloned from halophyte *Salicornia brachiata* confers salt and drought stress tolerance in transgenic tobacco. Marine biotechnology, 16, pp.321-332.

Sornaraj, P., Luang, S., Lopato, S. and Hrmova, M., 2016. Basic leucine zipper (bZIP) transcription factors involved in abiotic stresses: a molecular model of a wheat bZIP factor and implications of its structure in function. Biochimica et Biophysica Acta (BBA)-General Subjects, 1860(1), pp.46-56.

Tena, G., Asai, T., Chiu, W.L. and Sheen, J., 2001. Plant mitogen-activated protein kinase signaling cascades. Current opinion in plant biology, 4(5), pp.392-400.

Tiwari, R.K., Lal, M.K., Kumar, R., Chourasia, K.N., Naga, K.C., Kumar, D., Das, S.K. and Zinta, G., 2021. Mechanistic insights on melatonin-mediated drought stress mitigation in plants. Physiologia Plantarum, 172(2), pp.1212-1226.

Tousoulis, D., Kampoli, A.M., Tentolouris Nikolaos Papageorgiou, C. and Stefanadis, C., 2012. The role of nitric oxide on endothelial function. Current vascular pharmacology, 10(1), pp.4-18.

Tripathi, P., Rabara, R.C. and Rushton, P.J., 2014. A systems biology perspective on the role of WRKY transcription factors in drought responses in plants. Planta, 239, pp.255-266.

Tu, M., Wang, X., Yin, W., Wang, Y., Li, Y., Zhang, G., Li, Z., Song, J. and Wang, X., 2020. Grapevine VlbZIP30 improves drought resistance by directly activating VvNAC17 and promoting lignin biosynthesis through the regulation of three peroxidase genes. Horticulture research, 7.

Tweneboah, S. and Oh, S.K., 2017. Biological roles of NAC transcription factors in the regulation of biotic and abiotic stress responses in solanaceous crops. Journal of Plant Biotechnology, 44(1), pp.1-11.

Vlot, A.C., Dempsey, D.M.A. and Klessig, D.F., 2009. Salicylic acid, a multifaceted hormone to combat disease. Annual review of phytopathology, 47, pp.177-206.

Wang, B., Song, N., Zhang, Q., Wang, N. and Kang, Z., 2018. TaMAPK4 acts as a positive regulator in defense of wheat stripe-rust infection. Frontiers in Plant Science, 9, p.152.

Wang, W., Vinocur, B., Shoseyov, O. and Altman, A., 2004. Role of plant heat-shock proteins and molecular chaperones in the abiotic stress response. Trends in plant science, 9(5), pp.244-252.

Wang, Y., Yu, Q., Li, Y., Li, J., Chen, J., Liu, Z., Huang, J.A., Al-Harbi, M.S., Ali, E.F. and Eissa, M.A., 2021. Mechanisms of nitric oxide in the regulation of chilling stress tolerance in *Camellia sinensis*. Horticulturae, 7(10), p.410.

Wang, Y., Zhou, L.J., Wang, Y., Geng, Z., Liu, S., Chen, C., Chen, S., Jiang, J. and Chen, F., 2022. CmMYB9a activates floral coloration by positively regulating anthocyanin biosynthesis in chrysanthemum. Plant Molecular Biology, pp.1-13.

Wei, X., Shan, T., Hong, Y., Xu, H., Liu, X. and Zhang, Z., 2017. TaPIMP2, a pathogen-induced MYB protein in wheat, contributes to host resistance to common root rot caused by Bipolaris sorokiniana. Scientific reports, 7(1), p.1754.

Wilson, I.D., Neill, S.J. and Hancock, J.T., 2008. Nitric oxide synthesis and signaling in plants. Plant, cell & environment, 31(5), pp.622-631.

Windram, O., Penfold, C.A. and Denby, K.J., 2014. Network modeling to understand plant immunity. Annual review of phytopathology, 52, pp.93-111.

Xu, B., Ohtani, M., Yamaguchi, M., Toyooka, K., Wakazaki, M., Sato, M., Kubo, M., Nakano, Y., Sano, R., Hiwatashi, Y. and Murata, T., 2014. Contribution of NAC transcription factors to plant adaptation to land. Science, 343(6178), pp.1505-1508.

Xu, J., Yin, H., Li, Y. and Liu, X., 2010. Nitric oxide is associated with long-term zinc tolerance in Solanum nigrum. Plant Physiology, 154(3), pp.1319-1334.

Xuan, W., Xu, S., Li, M., Han, B., Zhang, B., Zhang, J., Lin, Y., Huang, J., Shen, W. and Cui, J., 2012. Nitric oxide is involved in hemin-induced cucumber adventitious rooting process. Journal of plant physiology, 169(11), pp.1032-1039.

Yang, D.L., Yang, Y. and He, Z., 2013. Roles of plant hormones and their interplay in rice immunity. Molecular plant, 6(3), pp.675-685.

Yin, Z., Wang, J., Wang, D., Fan, W., Wang, S. and Ye, W., 2013. The MAPKKK gene family in Gossypium raimondii: genome-wide identification, classification and expression analysis. International journal of molecular sciences, 14(9), pp.18740-18757.

Yuan, X., Huang, P., Wang, R., Li, H., Lv, X., Duan, M., Tang, H., Zhang, H. and Huang, J., 2018. A zinc finger transcriptional repressor confers pleiotropic effects on rice growth and drought tolerance by down-regulating stress-responsive genes. Plant and Cell Physiology, 59(10), pp.2129-2142.

Zhang, L., Li, Y., Lu, W., Meng, F., Wu, C.A. and Guo, X., 2012. Cotton GhMKK5 affects disease resistance, induces HR-like cell death, and reduces the tolerance to salt and drought stress in transgenic Nicotiana benthamiana. Journal of experimental botany, 63(10), pp.3935-3951.

Zhang, Y., He, J., Xiao, Y., Zhang, Y., Liu, Y., Wan, S., Liu, L., Dong, Y., Liu, H. and Yu, Y., 2021. CsGSTU8, a glutathione S-transferase from *Camellia sinensis*, is regulated by CsWRKY48 and plays a positive role in drought tolerance. Frontiers in Plant Science, 12, p.795919.

Zhang, Y., Xia, G. and Zhu, Q., 2021. Conserved and unique roles of chaperone-dependent E3 ubiquitin ligase CHIP in plants. Frontiers in plant science, 12, p.699756.

Zhang, Y., Yang, C., Li, Y., Zheng, N., Chen, H., Zhao, Q., Gao, T., Guo, H. and Xie, Q., 2007. SDIR1 is a RING finger E3 ligase that positively regulates stress-responsive abscisic acid signaling in Arabidopsis. The Plant Cell, 19(6), pp.1912-1929.

Zhao, J., 2007. Interplay among nitric oxide and reactive oxygen species: a complex network determining cell survival or death. Plant signaling & behavior, 2(6), pp.544-547.

Zhao, L., Zhang, F., Guo, J., Yang, Y., Li, B. and Zhang, L., 2004. Nitric oxide functions as a signal in salt resistance in the calluses from two ecotypes of reed. Plant Physiology, 134(2), pp.849-857.

Zhao, M.M., Zhang, X.W., Liu, Y.W., Li, K., Tan, Q., Zhou, S., Wang, G. and Zhou, C.J., 2020. A WRKY transcription factor, TaWRKY42-B, facilitates initiation of leaf senescence by promoting jasmonic acid biosynthesis. BMC Plant Biology, 20(1), pp.1-22.

Zhou, J.M., Trifa, Y., Silva, H., Pontier, D., Lam, E., Shah, J. and Klessig, D.F., 2000. NPR1 differentially interacts with members of the TGA/OBF family of transcription factors that bind an element of the PR-1 gene required for induction by salicylic acid. Molecular Plant-Microbe Interactions, 13(2), pp.191-202.

# CHAPTER 7

# SALICYLIC ACID AND NITRIC OXIDE: HEAT STRESS MEMORY THROUGH EPIGENETIC MODIFICATIONS

## Abstract

Plants possess complex defense mechanisms to survive and adapt to environmental stresses. Stress memory, the ability of plants to retain a memory of previous stress events, plays a vital role in enhancing their defense responses. Salicylic acid (SA) and nitric oxide (NO) are essential signaling molecules in stress memory development. SA activates defense genes, enhances antimicrobial compound production, and orchestrates systemic acquired resistance. It also regulates the priming of defense responses, preparing plants for future stress encounters. NO acts as a mediator in local and systemic defense responses and contributes to transgenerational heat stress memory. The interplay between SA and NO in heat stress memory involves their production during heat stress, their interaction in regulating heat stress-responsive genes, and their synergistic effects on defense mechanisms. Furthermore, SA and NO influence epigenetic modifications (Figure 1), including DNA methylation and histone modifications, which are involved in heat stress memory. Understanding the molecular mechanisms and interplay between SA, NO, and epigenetic changes is crucial for improving plant resilience to stress conditions.

**Keywords**: Heat stress, Epigenetic, Stress memory, Small RNAs

## Introduction

Plants are constantly exposed to various environmental stresses such as drought, high temperatures, pathogens, and pollutants. Plants have evolved complex defense mechanisms to survive and adapt to these challenging conditions. Among these defense mechanisms, stress memory is crucial in enabling plants to respond more effectively to recurrent stress episodes.

Stress memory refers to the phenomenon by which plants retain a "memory" of previous stress events and can mount a faster and more robust defense response upon subsequent encounters with the same stressor (Hilker et al., 2016). SA and NO are essential signaling molecules in stress memory. In this article, we will explore the role of SA and NO in the development of stress memory in plants. Salicylic acid is a naturally occurring phytohormone involved in various physiological processes, including plant defense responses. It plays a vital role in activating defense genes, enhancing the production of antimicrobial compounds, and orchestrating systemic acquired resistance (SAR), a long-lasting defense mechanism that protects not only the localized site of infection but also distant parts of the plant (Dempsey et al., 2011). SA is synthesized in response to stress signals, such as pathogen attack or abiotic stress, and is a critical regulator in stress memory development (Beckers and Spoel, 2005). Nitric oxide, on the other hand, is a gaseous signaling molecule that has been implicated in numerous physiological processes in plants, including plant growth, development, and stress responses (Correa-Aragunde et al., 2004). Nitric oxide participates in stress memory by modulating various cellular processes, such as stomatal closure, antioxidant defense, and gene expression regulation (Neill et al., 2002). It acts as a secondary messenger in stress signaling pathways and interacts with other signaling molecules, including SA, to regulate stress memory development in plants. The interplay between SA and NO in stress memory is complex and involves intricate signaling networks. Both molecules can positively or negatively regulate each other's levels and signaling pathways, depending on the specific stress context (Rai et al., 2020). Understanding the precise mechanisms underlying the crosstalk between SA and NO in stress memory will provide valuable insights into how plants adapt and respond to environmental challenges.

During the initial stress exposure, SA levels increase rapidly, triggering a cascade of molecular events that activate defense responses. These responses involve expressing stress-related genes, synthesizing pathogenesis-related (PR) proteins, and reinforcing the plant cell wall (Klessig et al., 2018; Vlot et al., 2009). Importantly, SA also regulates the priming of defense responses, a process by which plants prepare themselves for future stress encounters. Priming involves modifying the expression of defense-related genes, reprogramming metabolic pathways, and reallocating resources to facilitate a more rapid and efficient response upon subsequent stress exposure (Martinez-Medina et al., 2016; Slaughter et al., 2012). NO is a gaseous signaling molecule versatile in plant physiology, including defense responses. It is involved in regulating plant growth, development, and stress adaptation. NO is synthesized enzymatically by nitric oxide synthase-like (NOS-like) enzymes or as a byproduct of nitrate reductase activity (Neill et al., 2002; Corpas and Barroso, 2017). In stress memory, NO is a crucial mediator in local and systemic defense responses (Yun et al., 2011; Rudolf et al., 2021).

Upon stress perception, NO levels increase rapidly in plants, activating various defense mechanisms. NO participates in the modulation of ion channels, protein kinases, and transcription factors, thereby regulating the expression of stress-responsive genes (Shapiro 2005; Corpas et al., 2017). In stress memory development, NO acts as a signal molecule that facilitates the transmission of defense signals from the initial stress site to distant plant parts. This systemic signaling allows the priming and activation of defense responses in the unexposed tissues, enabling a faster and stronger defense reaction during subsequent stress encounters (Feng et al., 2018; Asgher et al., 2017). SA and NO not only play crucial roles in the immediate defense responses of plants but also contribute to the development of stress memory through epigenetic regulation. Epigenetic modifications involve changes in gene expression without alterations in the DNA sequence, and they can be inherited from generation to generation. SA and NO have been found to influence the epigenetic landscape of plants, leading to long-lasting changes in gene expression patterns associated with stress memory. These modifications include DNA methylation, histone modifications, and the activity of small RNAs (Song et al., 2008; Pfluger and Wagner, 2007;). By modulating the epigenetic marks on genes involved in stress responses, SA and NO contribute to the establishment of stress memory, allowing plants to respond more effectively to recurring stress events. Therefore, this chapter will unravel the specific mechanisms by which SA and NO mediate epigenetic regulation and explore the potential applications of targeting epigenetic processes for enhancing plant stress tolerance.

## Epigenetic Modifications in Heat Stress Memory

Epigenetic modifications in heat stress memory refer to the changes in gene expression and cellular responses that occur in plants due to previous exposure to high temperatures. These modifications involve alterations to the epigenetic landscape, such as DNA methylation and histone modifications, which can persist even after the stress has been removed (Paluvai et al., 2020). When plants are exposed to heat stress, they undergo physiological and biochemical changes to cope with the unfavorable conditions. Remarkably, the plant can remember some of these changes, leading to enhanced tolerance and improved responses upon subsequent heat stress events (Table 1). This phenomenon is known as heat stress memory (Lämke and Bäurle, 2017; Shekhawat et al., 2022).

Epigenetic modifications play a crucial role in heat stress memory. Studies have shown that heat stress can induce changes in DNA methylation patterns, leading to altered gene expression in response to future heat stress events (Pick et al., 2012; Ding et al., 2014). Specific genes involved in stress response pathways, such as those related to heat shock proteins, transcription factors, and antioxidants, can be epigenetically regulated to facilitate a faster and more efficient response to heat stress (Zhang et al., 2007). In addition to DNA methylation, histone modifications contribute to heat stress memory. Histone acetylation, methylation, and phosphorylation have been implicated in the regulation of heat stress-responsive genes and the establishment of heat stress memory (Kim et al., 2010). These modifications affect chromatin structure and accessibility, influencing gene expression patterns under heat stress conditions. Understanding the epigenetic changes in heat stress memory is crucial for developing strategies to enhance plant resilience to high temperatures. By deciphering the underlying mechanisms and identifying vital epigenetic regulators, researchers can potentially manipulate these modifications to improve crop tolerance to heat stress, ultimately ensuring food security and agricultural sustainability in climate change.

**Table 1** Role of salicylic acid (SA) and nitric oxide (NO) in regulating epigenetic stress memory in plants

| **Role of SA and NO in Regulating Epigenetic Stress Memory** | **Epigenetic Modifier** | **Function** | **SA-Mediated Regulation** | **NO-Mediated Regulation** | **References** |
|---|---|---|---|---|---|
| **DNA Methylation** | DNA Methyltransferase | Addition of a methyl group to cytosine residues, leading to DNA methylation | SA induces the expression and activity of DNA methyltransferases, resulting in increased DNA methylation | NO modulates the activity of DNA methyltransferases, influencing DNA methylation levels | Dowen et al. (2012) |
| **Histone Modifications** | Histone Acetyltransferase (HAT) | Adds acetyl groups to histone proteins, leading to relaxed chromatin structure and increased gene expression | SA upregulates the expression and activity of HATs, promoting histone acetylation and gene activation | NO modulates the activity of HATs, influencing histone acetylation levels and gene expression | Engler and Jakob, (2013) |

| | | | | | |
|---|---|---|---|---|---|
| | Histone Deacetylase (HDAC) | Removes acetyl groups from histone proteins, leading to condensed chromatin structure and decreased gene expression | SA inhibits the activity of HDACs, promoting histone acetylation and gene activation | NO enhances the activity of HDACs, influencing histone deacetylation levels and gene expression | Bai et al. (2012) |
| | Histone Methyltransferase | Adds methyl groups to histone proteins, resulting in either gene activation or gene repression | SA regulates the expression and activity of histone methyltransferases, modulating histone methylation patterns | NO modulates the activity of histone methyltransferases, influencing histone methylation levels and gene expression | Grima-Pettenati et al. (2012) |

## Definition and explanation of epigenetic modifications

Epigenetic modifications in plants refer to heritable gene expression or cellular phenotype changes without alterations to the underlying DNA sequence. These modifications involve chemical changes in the DNA molecule itself or changes to the proteins associated with DNA, known as histones. Epigenetic mechanisms are crucial in regulating gene expression, development, and responses to environmental cues in plants (Iwasaki and Paszkowski, 2014; Iwasaki and Paszkowski, 2014).

One of the most well-known epigenetic modifications is DNA methylation. DNA methylation involves adding a methyl group to the DNA molecule, usually at cytosine residues. In plants, DNA methylation occurs in specific regions called CpG islands, which are often found in gene promoter regions. Methylation of these regions can lead to gene silencing, preventing the transcription machinery from accessing the DNA and inhibiting gene expression. Conversely, the demethylation of DNA can activate gene expression (Law and Jacobsen, 2010; Rai et al., 2020a). Histone modifications are another crucial epigenetic mechanism in plants. Histones are proteins that help package DNA into a compact structure called chromatin. Different chemical modifications, such as acetylation, methylation, phosphorylation, and ubiquitination, can occur on histone proteins, altering their interaction with DNA and influencing gene expression. For example, acetylation of histones is generally associated with gene activation, while methylation can be linked to gene silencing or activation, depending on the specific residues and context (Springer, 2013; Otero et al., 2014).

Both genetic and environmental factors can influence epigenetic modifications in plants. Stress conditions, such as temperature extremes, drought, or nutrient deficiencies, can trigger changes in DNA methylation patterns or histone modifications, leading to altered gene expression and adaptive responses (Kooke et al., 2015; Latzel et al., 2016; Secco et al., 2013). Moreover, epigenetic modifications can be inherited through mitotic or meiotic cell divisions, allowing plants to pass on epigenetic information to their offspring (Johannes et al., 2009; Quadrana and Colot, 2016). Understanding and manipulating plant epigenetic changes have important implications for agriculture and crop improvement. Researchers can potentially develop strategies to enhance crop performance, disease resistance, and sustainability by deciphering the epigenetic code and its interaction with the environment (Iwasaki and Paszkowski, 2014; Zhang et al., 2007).

## Epigenetic mechanisms involved in plant heat stress memory

Epigenetic mechanisms play a significant role in plant heat stress memory, which refers to the ability of plants to retain a "memory" of previous heat stress events and mount a more efficient response upon subsequent exposures. These mechanisms involve changes in DNA methylation and histone modifications that persist even after removing the initial heat stress (Song et al., 2008; Wang and Li, 2006). One substantial epigenetic modification in heat stress memory is DNA methylation. Heat stress can induce alterations in the methylation patterns of specific genes, leading to changes in their expression levels (Rudolf et al., 2021). This altered DNA methylation pattern can be inherited through cell divisions and prime the plant for a more rapid and effective response to future heat stress (Seo et al., 2010; Li et al., 2019). It allows the plant to fine-tune the expression of stress-related genes and activate stress response pathways more efficiently.

**Table 2** molecular mechanisms of epigenetic modifications in heat stress memory and the role of salicylic acid (SA) and nitric oxide (NO) in stimulating post-translational modifications

| Molecular Mechanisms of Epigenetic Modifications in Heat Stress Memory | Epigenetic Modification | Function | Role of SA | Role of NO | References |
|---|---|---|---|---|---|
| **DNA Methylation** | DNA Methyltransferase | Catalyzes the addition of a methyl group to cytosine residues, leading to DNA methylation | SA enhances the expression and activity of DNA methyltransferases, resulting in increased DNA methylation | NO modulates the activity of DNA methyltransferases, influencing DNA methylation levels | Dowen et al. (2012) |
| **Histone Modifications** | Histone Acetylation | Addition of acetyl groups to histone proteins, leading to relaxed chromatin structure and increased gene expression | SA promotes histone acetylation through the activation of histone acetyltransferases | NO modulates the activity of histone acetyltransferases, affecting histone acetylation levels | Engler and Jakob, (2013) |

| | | | | | |
|---|---|---|---|---|---|
| | Histone Methylation | Addition of methyl groups to histone proteins, resulting in gene activation or repression | SA modulates the expression and activity of histone methyltransferases, influencing histone methylation patterns | NO regulates the activity of histone methyltransferases, affecting histone methylation levels | Kim et al. (2010) |
| | Histone Phosphorylation | Addition of phosphate groups to histone proteins, influencing chromatin remodeling and gene expression | SA induces the phosphorylation of histone proteins, affecting chromatin remodeling and gene expression | NO regulates the phosphorylation of histone proteins, modulating chromatin remodeling and gene expression | Qiao and Fan, (2008) |
| **Non-coding RNA** | microRNAs (miRNAs) | Small RNA molecules that regulate gene expression by targeting messenger RNAs (mRNAs) for degradation or translational repression | SA and NO regulate the expression and activity of miRNAs, affecting the post-transcriptional regulation of target genes | SA and NO modulate the expression and activity of miRNAs, influencing the post-transcriptional regulation of target genes | Sunkar et al. (2007) |

| | | | | | |
|---|---|---|---|---|---|
| | long non-coding RNAs (lncRNAs) | Longer RNA molecules with diverse regulatory roles in gene expression and chromatin organization | SA and NO regulate the expression and activity of lncRNAs, modulating their interactions with chromatin and gene regulation | SA and NO modulate the expression and activity of lncRNAs, influencing their interactions with chromatin and gene regulation | Mammarella et al. (2023) |
| **Protein Modifications** | Phosphorylation | Addition of phosphate groups to proteins, altering their activity and function | SA stimulates the phosphorylation of proteins involved in heat stress memory | NO regulates the phosphorylation of proteins involved in heat stress memory | Essemine et al. (2010) |
| | Acetylation | Addition of acetyl groups to proteins, influencing their function and stability | SA enhances the acetylation of proteins involved in heat stress memory | NO modulates the acetylation of proteins involved in heat stress memory | Bhattacharjee (2012) |
| | Ubiquitination | Attachment of ubiquitin molecules to proteins, regulating their degradation or function | SA and NO regulate the ubiquitination of proteins involved in heat stress memory | SA and NO modulate the ubiquitination of proteins involved in heat stress memory | Sang et al. (2008). |

Histone modifications also play a role in plant heat stress memory. Heat stress can lead to changes in histone acetylation, methylation, phosphorylation, and other modifications, which influence the accessibility of DNA and gene expression (Table 2) (Singroha and Sharma, 2019; Pick et al., 2012). These modifications can create a more permissive chromatin structure that facilitates the activation of stress-responsive genes during subsequent heat stress events (Sequeira-Mendes et al., 2014; Lämke and Bäurle, 2017). Additionally, the presence of specific histone modifications can act as marks or signals for the recruitment of proteins involved in stress response pathways (Ding et al., 2021; Rai et al., 2020). By understanding and manipulating these epigenetic mechanisms, researchers aim to improve the heat stress tolerance of crops. Targeting specific epigenetic regulators or modifying the epigenetic landscape of plants can enhance their ability to remember past heat stress events, resulting in improved resilience and productivity under high-temperature conditions (Bauer et al., 2015; Lämke and Bäurle, 2017).

## Importance of epigenetic modifications for transgenerational heat stress memory

Epigenetic modifications play a crucial role in transgenerational heat stress memory, which refers to the phenomenon where the effects of heat stress experienced by parents are passed on to their offspring, resulting in enhanced tolerance to heat stress. These modifications are essential for transmitting and perpetuating the acquired traits across generations, ensuring the survival and adaptation of plants in changing environments. DNA methylation is one of the key epigenetic modifications in transgenerational heat stress memory. When plants are exposed to heat stress, DNA methylation patterns can change in specific genome regions. The next generation can inherit these alterations through somatic and germ cells, leading to altered gene expression patterns. The transgenerational transmission of these DNA methylation marks can prime the offspring for improved heat stress tolerance by activating stress-responsive genes and fine-tuning their expression (Iwasaki and Paszkowski, 2014; Hu et al., 2014; Garg et al., 2015).

Histone modifications also contribute to transgenerational heat stress memory. Heat stress can induce changes in histone acetylation, methylation, and other modifications that can be inherited across generations. These modifications can affect the accessibility of genes involved in stress responses and regulate their expression levels. The transgenerational transmission of altered histone modifications can provide a "memory" of the heat stress experience and enable the offspring to respond more effectively to subsequent heat stress events. The importance of epigenetic modifications for transgenerational heat stress memory lies in their ability to confer enhanced stress tolerance to offspring without alterations in the underlying DNA sequence. This mechanism allows plants to adapt and survive in rapidly changing environments by transmitting acquired traits to future generations. The offspring are better equipped to cope with heat stress conditions by inheriting the epigenetic modifications associated with heat stress memory, resulting in improved growth, survival, and reproductive success (Chinnusamy and Zhu, 2009; Bauer et al., 2015; Byrne, 2012).

Understanding the mechanisms underlying transgenerational heat stress memory and the specific epigenetic modifications involved is essential for developing strategies to enhance crop resilience. Manipulating these modifications through breeding or genetic engineering approaches can potentially improve the heat stress tolerance of crops, ensuring sustainable food production in the face of climate change. Epigenetic modifications provide a valuable avenue for transmitting acquired traits and facilitating rapid adaptation to changing environmental conditions, contributing to plant populations' long-term survival and success (Paluvai et al., 2020; Rai et al., 2020).

## Salicylic Acid: Modulating Epigenetic Responses to Heat Stress

SA is a naturally occurring plant hormone that plays a vital role in plant defense mechanisms against various stresses, including heat stress (Miura & Tada, 2014; Rivas-San Vicente & Plasencia, 2011). Heat stress is a significant environmental challenge that negatively impacts plant growth, development, and crop productivity. However, recent research has revealed that SA can modulate epigenetic responses to heat stress, offering a potential avenue for enhancing heat stress tolerance in plants (Rai et al., 2018). Epigenetic modifications are heritable changes in gene expression patterns without alterations to the underlying DNA sequence. They play a crucial role in regulating gene expression and can be influenced by environmental factors, including heat stress. SA has been shown to regulate the expression of genes involved in heat stress response by modulating epigenetic modifications such as DNA methylation and histone modifications (Rai et al., 2018).

Studies have demonstrated that SA treatment can alleviate heat stress-induced DNA hypermethylation associated with the repression of heat stress-responsive genes (Chen, 2021; Khanna et al., 2023). SA can also influence histone modifications, including histone acetylation and methylation, which affect gene expression (Jaskiewicz et al., 2011; Růžička et al., 2017). By modulating these epigenetic marks, SA can enhance the presentation of heat stress-responsive genes, improving plants' heat stress tolerance (Gao et al., 2022; Kovacs et al., 2015). Furthermore, SA-mediated epigenetic changes can have transgenerational effects, meaning that the enhanced heat stress tolerance conferred by SA treatment can be passed on to subsequent generations (Matzke et al., 2009). This phenomenon highlights the potential of SA as a promising tool for breeding heat-stress-tolerant crops

## Salicylic acid-mediated epigenetic modifications under heat stress

SA is a plant hormone extensively studied for its role in plant defense mechanisms against various stresses, including heat stress (Miura and Tada, 2014; Khan et al., 2015). Heat stress poses a significant environmental challenge that can harm plant growth, development, and crop productivity (Wahid et al., 2007). Recent research has shed light on the involvement of SA in mediating epigenetic modifications under heat stress, providing insights into the molecular mechanisms underlying heat stress response in plants (Kovacs et al., 2015; Pick et al., 2012). Epigenetic modifications refer to heritable changes in gene expression that do not involve alterations to the DNA sequence (Hauben et al., 2009). Environmental factors, including heat stress, can influence these modifications (Hu et al., 2014). SA has been found to play a crucial role in modulating epigenetic responses to heat stress, thereby regulating gene expression patterns that contribute to heat stress tolerance in plants (Rivas-San Vicente and Plasencia, 2011; Shu et al., 2018).

DNA methylation is one of the key epigenetic modifications influenced by SA under heat stress (Feng et al., 2018; Rai et al., 2020). DNA methylation involves adding a methyl group to the DNA molecule, which can affect gene expression (Law and Jacobsen, 2010). Heat stress often leads to DNA hypermethylation, resulting in the repression of heat stress-responsive genes (Ding et al., 2012; Cong et al., 2019). However, SA treatment has been shown to counteract this hypermethylation by promoting DNA demethylation, thereby alleviating the repression of heat stress-responsive genes (Ding et al., 2012; Rai et al., 2018, 2020). In addition to DNA methylation, SA can also influence histone modifications, which are chemical changes to the proteins around which DNA is wound (Kim et al., 2010). Histone modifications, such as acetylation and methylation, can affect the accessibility of genes to the transcriptional machinery, thereby regulating gene expression (Otero et al., 2014). SA treatment has been found to modulate histone acetylation and methylation patterns under heat stress conditions, leading to the activation of heat stress-responsive genes (Rai et al., 2018; He et al., 2022).

Furthermore, SA-mediated epigenetic modifications can have transgenerational effects, meaning that subsequent generations of plants can inherit the changes induced by SA treatment (Cong et al., 2019; Rai et al., 2018). This transgenerational memory of heat stress response can contribute to offspring plants' enhanced heat stress tolerance (Ding et al., 2009; Wang and Li, 2006). Understanding the role of SA in mediating epigenetic modifications under heat stress is essential for developing strategies to strengthen heat stress tolerance in crops. By unraveling the molecular mechanisms, researchers can explore targeted approaches to manipulating epigenetic marks and gene expression patterns to improve plant resilience to heat stress (Song et al., 2012).

## Role of salicylic acid in transgenerational heat stress memory

SA is a plant hormone that has gained attention for its involvement in various physiological processes, including plant defense against abiotic stresses like heat stress (Miura and Tada, 2014). Heat stress is a significant environmental challenge that can have long-lasting effects on plant growth and development (Bita and Gerats, 2013). Recent studies have shown that SA plays a role in transgenerational heat stress memory, wherein the impact of heat stress experienced by parent plants is transmitted to their offspring, resulting in enhanced heat stress tolerance (Ding et al., 2009). Transgenerational memory refers to the ability of plants to inherit information from their parent generation about past environmental conditions and use it to better adapt to similar stresses in subsequent generations. SA is pivotal in establishing this transgenerational memory in response to heat stress (Cong et al., 2019; Wang and Li, 2006)

When parent plants are exposed to heat stress, they accumulate SA, which acts as a signaling molecule and triggers a cascade of molecular events

(Paluvai et al., 2020). SA-induced changes in gene expression, DNA methylation patterns, and chromatin modifications help the parent plants adapt and respond to heat stress (Munir et al., 2018). Remarkably, these changes can be inherited by the offspring plants through epigenetic inheritance (Iwasaki and Paszkowski, 2014). Epigenetic modifications, such as DNA methylation and histone modifications, are crucial in transgenerational memory (Shanker et al., 2020). Heat stress-induced DNA methylation patterns can be maintained and transmitted to the next generation, influencing gene expression patterns in the offspring plants (Ding et al., 2012). SA has been shown to be involved in establishing and maintaining these DNA methylation marks, ensuring the transmission of heat stress memory (Rai et al., 2021).

Furthermore, SA can influence histone modifications, which can affect the accessibility of genes to the transcriptional machinery. Heat stress-induced alterations in histone modifications can also be transmitted to the next generation, providing a molecular basis for transgenerational memory (Pick et al., 2012). SA has been implicated in regulating these histone modifications and ensuring their transgenerational inheritance (Rai et al., 2021). The role of SA in transgenerational heat stress memory highlights its importance in promoting the adaptation and survival of plants under recurring heat stress conditions (Chen, 2021). Plants can better cope with future heat stress events by transmitting information about past heat stress experiences to subsequent generations. Understanding the mechanisms underlying SA-mediated transgenerational memory will provide valuable insights into plant resilience and may have implications for improving crop performance under changing climate conditions. Overall, the involvement of SA in transgenerational heat stress memory highlights its significance as a critical regulator in plant responses to stress. It also opens up avenues for further research on epigenetic inheritance and the inheritance of acquired traits in plants.

## Nitric Oxide: Epigenetic Regulation in Heat Stress Response

Recent research has revealed that NO is involved in the epigenetic regulation of gene expression during heat stress response in plants (Prakash et al., 2021). Epigenetic modifications are heritable changes in gene expression patterns that occur without altering the underlying DNA sequence. These modifications, including DNA methylation and histone modifications, are vital in regulating gene expression (Chinnusamy and

Zhu 2009; Zhou et al., 2015). Under heat stress conditions, NO has been shown to influence the activity of enzymes involved in DNA methylation and histone modifications, thereby modulating gene expression (Nabi et al., 2019; Rai et al., 2018). NO has been found to affect DNA methylation patterns during heat stress. It can regulate the activity of DNA methyltransferases, enzymes responsible for adding methyl groups to DNA molecules (Cong et al., 2019). By modulating DNA methylation, NO can control the expression of genes involved in heat stress response and tolerance (Rai et al., 2018).

Furthermore, NO can also influence histone modifications, which affect the accessibility of genes to the transcriptional machinery. NO has been shown to regulate the activity of enzymes responsible for histone modifications, such as histone acetylation and methylation (Kohli et al., 2022; Mengel et al., 2017). These modifications can alter the chromatin structure and impact gene expression patterns during heat stress. The epigenetic regulation mediated by NO in the heat stress response provides plants with a mechanism to modulate gene expression rapidly and adapt to changing environmental conditions (Kohli et al., 2022; Mengel et al., 2017). By influencing DNA methylation and histone modifications, NO can fine-tune the expression of heat stress-responsive genes and enhance plant tolerance to heat stress (Rai et al., 2021; Asgher et al., 2017).

Understanding the role of NO in epigenetic regulation during the heat stress response is vital for unraveling the molecular mechanisms underlying plant adaptation to heat stress. Further research in this area will contribute to a better understanding of how plants utilize NO-mediated epigenetic changes to cope with heat stress and may have implications for developing strategies to improve crop performance under high-temperature conditions.

## Nitric oxide-induced epigenetic changes under heat stress

NO is a gaseous signaling molecule that has emerged as a critical player in plant responses to heat stress (Rai et al., 2020, 2021). Recent studies have shed light on the involvement of NO in inducing epigenetic changes, which are heritable alterations in gene expression patterns that do not involve changes to the DNA sequence during heat stress (Korotko et al., 2021; Rai et al., 2020). Under heat-stress conditions, plants produce NO as part of their defense mechanisms (Lee et al., 2014). NO acts as a signaling molecule that triggers a cascade of molecular events, including modulating epigenetic modifications. DNA methylation is one of the major epigenetic changes influenced by NO under heat stress (Rai et al., 2021; Rai et al., 2020).

DNA methylation involves adding a methyl group to the DNA molecule, which can affect gene expression. Heat stress often leads to changes in DNA methylation patterns, and NO has been shown to play a role in regulating this process (Paluvai et al., 2020; He et al., 2022). NO can induce or inhibit DNA methylation, depending on the genes involved. These alterations in DNA methylation patterns can lead to changes in gene expression, allowing plants to respond and adapt to heat stress (Hu et al., 2014; Parankusam et al., 2017). NO can also impact histone modifications, which are chemical changes to the proteins around which DNA is wrapped. Histone modifications can alter the accessibility of genes to the transcriptional machinery and influence gene expression. NO has been found to regulate the activity of enzymes responsible for histone modifications under heat stress conditions (Rai et al., 2020; Rai et al., 2018). This modulation of histone modifications can further fine-tune gene expression patterns and contribute to the plant's heat stress response (Hu et al., 2014; Parankusam et al., 2017).

The NO-induced epigenetic changes observed during heat stress are crucial for the plant's ability to adapt and survive under these challenging conditions (Rai et al., 2020a; He et al., 2022). By altering DNA methylation and histone modifications, NO can regulate gene expression in heat stress response and tolerance (Kovacs et al., 2015). Understanding the role of NO-induced epigenetic changes in heat stress response is essential for developing strategies to enhance plant resilience to high temperatures (Hu et al., 2014; Rai et al., 2018). By unraveling the molecular mechanisms underlying these epigenetic modifications, researchers can explore targeted approaches to manipulate gene expression patterns and improve heat stress tolerance in crops (Rai et al., 2021; Li et al., 2019). Further studies in this field will provide valuable insights into the interplay between NO signaling, epigenetic changes, and the plant's response to heat stress.

## Nitric Oxide's Contribution to transgenerational heat stress memory

NO has been recognized as a crucial signaling molecule in plant responses to heat stress. Recent studies have shed light on the role of NO in contributing to transgenerational heat stress memory, which refers to the phenomenon where plants inherit information about heat stress experienced by their parent generation and exhibit enhanced tolerance to subsequent heat stress events (Wendehenne et al., 2014; Rai et al., 2020). During heat stress, plants produce NO as part of their defense mechanisms. This NO accumulation triggers a series of molecular events that enable plants to cope with the immediate heat stress and influence the transmission of heat stress memory to future generations (Prakash et al., 2021; Sharma et al., 2022).

NO contributes to transgenerational heat stress memory through its involvement in epigenetic modifications. Epigenetic changes, such as DNA methylation and histone modifications, can be inherited, impacting gene expression patterns in subsequent generations (Rai et al., 2020; He et al., 2022). NO has been found to influence these epigenetic marks during heat stress, leading to alterations in gene expression that persist across generations (Lee et al., 2014; Parankusam et al., 2017). NO can affect DNA methylation patterns by modulating the activity of enzymes involved in DNA methylation processes (Hu et al., 2014). This changes the methylation status of specific genes, which can impact their expression and contribute to the transgenerational transmission of heat stress memory.

Additionally, NO can also influence histone modifications, which regulate gene expression by modifying the accessibility of genes to the transcriptional machinery. NO has been shown to affect the activity of enzymes responsible for histone modifications, thereby altering gene expression patterns associated with heat stress response and memory (Wang et al., 2015; Rai et al., 2020). The involvement of NO in transgenerational heat stress memory highlights its importance as a molecular mediator in transmitting information about past heat stress experiences to subsequent generations (Zhou et al., 2012). By modulating epigenetic modifications, NO helps plants establish a heritable memory of heat stress, enhancing tolerance in future generations. Understanding the mechanisms underlying NO's contribution to transgenerational heat stress memory is significant for developing strategies to improve crop resilience (Wang et al., 2015). By manipulating NO signaling or targeting specific epigenetic pathways influenced by NO, it may be possible to enhance the transmission of heat stress memory and improve plant performance under high-temperature conditions.

## The Interplay between Salicylic Acid and Nitric Oxide in Heat Stress Memory

The interplay between SA and NO plays a crucial role in heat stress memory. When exposed to high temperatures, plants experience heat stress, which can harm their growth and development. However, plants have evolved mechanisms to cope with such pressure. Salicylic acid is a signaling molecule involved in various physiological processes in plants, including defense responses against pathogens and abiotic stresses (Khan et al., 2015). Nitric oxide, on the other hand, is a gaseous signaling molecule that also plays a role in plant stress responses (Shapiro 2005).

Studies have shown that heat stress induces the production of both SA and NO in plants. These two molecules interact to regulate the expression of heat stress-responsive genes, activating heat stress memory (Cong et al., 2019; Prakash et al., 2021). Heat stress memory allows plants to better tolerate subsequent heat stress episodes by priming their defense mechanisms. Overall, the interplay between SA and NO in heat stress memory highlights the intricate signaling network in plants that enables them to adapt and survive under challenging environmental conditions.

## Cross-talk and synergistic effects of salicylic acid and nitric oxide signaling

SA and NO are two essential signaling molecules in plants that play crucial roles in various physiological processes, including defense responses against pathogens and abiotic stresses (Miura and Tada, 2014). The cross-talk and synergistic effects between SA and NO signaling pathways have been extensively studied and are known to be intricately connected. Cross-talk refers to the communication and interaction between different signaling pathways. Studies have shown that these two molecules can modulate each other's levels and activities in the case of SA and NO (Klessig et al., 1998; Rai et al., 2020). For example, SA can enhance NO production in plants, while NO can induce the accumulation of SA. This cross-talk allows for a coordinated response to stress conditions.

The synergistic effects of SA and NO signaling pathways refer to their collective action, leading to enhanced defense responses. It has been observed that the simultaneous application of SA and NO can result in more excellent induction of defense genes compared to their application (Fritsche et al., 2017; Beligni and Lamattina, 2001). This suggests that SA and NO act synergistically to amplify the plant's defense mechanisms against stresses. Furthermore, SA and NO signaling pathways can converge on common downstream targets, such as transcription factors and enzymes involved in defense responses (Liu and Lang, 2020; Arasimowicz and Floryszak-Wieczorek, 2007). This convergence allows for an integrated and potentiated response to stress conditions.

Overall, the cross-talk and synergistic effects between SA and NO signaling pathways highlight the complexity and interconnectivity of plant signaling networks. These interactions play a vital role in fine-tuning plant defense responses and enabling plants to adapt and survive under challenging environmental conditions. Understanding the interplay between SA and NO signaling pathways can provide valuable insights into the mechanisms underlying plant stress responses. It can potentially be exploited to develop strategies to enhance crop stress tolerance

## Molecular Mechanisms of Epigenetic Modifications in Heat Stress Memory

The molecular mechanisms of epigenetic modifications in heat stress memory involve several vital processes. Heat stress triggers changes in DNA methylation patterns, histone modifications, and small RNA profiles, which can influence gene expression (Song et al., 2008; Queitsch et al., 2000). DNA methylation, controlled by DNA methyltransferases, can be altered by heat stress, leading to changes in gene activity (Dowen et al., 2012; Ding et al., 2009). Histone modifications, such as acetylation and methylation, can also be affected by heat stress, impacting the accessibility of DNA to transcriptional machinery (Yuan et al., 2013; Hu et al., 2014). Small RNAs, including microRNAs and small interfering RNAs, can regulate gene expression by targeting specific mRNA molecules (Liu et al., 2007; Feng et al., 2018). These epigenetic modifications serve as molecular marks that enable the transmission of heat stress memory to subsequent generations. Understanding the molecular mechanisms underlying these epigenetic modifications provides insights into the long-term effects of heat stress on gene regulation and the adaptive responses of plants to challenging environments. By unraveling the complex interplay between DNA methylation, histone modifications, and small RNA pathways, researchers can gain a deeper understanding of how plants establish and maintain heat stress memory, contributing to their resilience and survival.

## DNA methylation changes and their role in heat stress memory

DNA methylation changes play a crucial role in heat stress memory, which refers to the ability of plants to retain a "memory" of prior heat stress exposure and mount a more efficient response upon subsequent encounters with high temperatures (Chinnusamy and Zhu, 2009). DNA methylation is an epigenetic modification that involves adding a methyl group to the DNA molecule, typically occurring at cytosine residues in the context of a CpG dinucleotide (Feng et al., 2010; Law and Jacobsen, 2010). Heat stress can induce alterations in DNA methylation patterns, leading to changes in gene expression and cellular responses (Dowen et al., 2012; Ding et al., 2009). Studies have shown that heat stress can cause hypermethylation and hypomethylation of specific genomic regions.

Hypermethylation refers to increased DNA methylation levels, often occurring in gene promoter regions. Heat-induced hypermethylation can suppress gene expression, inhibiting the transcription of specific genes involved in stress responses (Dowen et al., 2012; Avramova, 2015). This helps plants conserve energy by reducing the expression of unnecessary or potentially harmful genes under heat-stress conditions. Conversely, hypomethylation refers to decreased DNA methylation levels, often occurring in transposable elements and repetitive sequences. Heat-induced hypomethylation can lead to the activation of transposable elements, which are mobile genetic elements that can disrupt gene function and stability (Ito et al., 2011; Lang-Mladek et al., 2010). This activation can trigger genome instability and stress-related genomic rearrangements, affecting plant adaptation and evolution.

The DNA methylation changes induced by heat stress can be involved in heat stress memory. Plants that have experienced heat stress can retain altered DNA methylation patterns even after the pressure subsides (Iwasaki and Paszkowski, 2014; Kinoshita and Seki, 2014). These changes can persist without the initial stressor and prime the plants for more efficient and rapid responses upon subsequent heat stress exposure. The role of DNA methylation in heat stress memory is multifaceted. It can regulate the expression of genes involved in heat stress responses, enabling a quicker and more robust response upon re-exposure to high temperatures (Dowen et al., 2012; Liu et al., 2018). Furthermore, the inheritance of altered DNA methylation patterns through the germline can transmit heat stress memory to the next generation, providing a form of adaptive memory that enhances the offspring's ability to cope with heat stress (Mirouze and Paszkowski, 2011; Hauser et al., 2011).

Mechanistically, the establishment and maintenance of DNA methylation patterns involve DNA methyltransferases, enzymes responsible for adding methyl groups to DNA (Liu and Lang, 2020). Heat stress can influence the activity and expression of these enzymes, leading to changes in DNA methylation patterns (Yong-Villalobos et al., 2015). Heat stress can also impact the activity of DNA demethylases, enzymes responsible for removing methyl groups from DNA, further contributing to the dynamic nature of DNA methylation changes (Brzezinka et al., 2016). DNA methylation changes play a critical role in heat stress memory by modulating gene expression, priming responses, and potentially transmitting adaptive traits to subsequent generations (Chinnusamy and Zhu, 2009; Iwasaki and Paszkowski, 2014). Elucidating the precise mechanisms underlying DNA methylation dynamics in heat stress memory will provide valuable insights into the molecular basis of plant adaptation to challenging environments, with potential applications for improving crop resilience to heat stress.

## Histone modifications and their impact on gene expression during heat stress

Histone modifications significantly regulate gene expression during heat stress, enabling plants to respond appropriately to challenging conditions. Histones are proteins around which DNA is wrapped, forming a structure called chromatin. Post-translational modifications of histones, such as acetylation, methylation, phosphorylation, and ubiquitination, can influence the accessibility of DNA and affect gene expression patterns (Berr et al., 2011; Liu et al., 2007; Rai et al., 2021). During heat stress, histone modifications dynamically change to modulate gene expression in response to environmental cues. One of the most studied histone modifications is histone acetylation. Acetylation of histones is generally associated with transcriptional activation, as it relaxes the chromatin structure, making genes more accessible to the transcriptional machinery (Paluvai et al., 2020; Kim et al., 2010). Heat stress can induce the acetylation of specific histones, promoting gene expression in stress responses. This allows plants to activate defense mechanisms and adapt to challenging conditions rapidly (Fragkostefanakis et al., 2016; Pérez-Salamó et al., 2014).

In addition to acetylation, histone methylation is another crucial modification that can impact gene expression during heat stress. Histone methylation can occur at different positions on histone tails, such as lysine and arginine residues, and can have diverse effects depending on the specific site and degree of methylation. For example, methylation of specific lysine residues can be associated with transcriptional activation (e.g., H3K4me3), while methylation at other sites can lead to transcriptional repression (e.g., H3K9me2/3) (Lu et al., 2023; Zhou et al., 2015). Heat stress can induce changes in the levels and patterns of histone methylation, thus modulating gene expression accordingly (Lämke et al., 2016; Zhang et al., 2007). Phosphorylation and ubiquitination of histones are additional modifications that can impact gene expression during heat stress. Phosphorylation of histones can affect chromatin structure and gene expression by altering the interactions between histones and other proteins (Paluvai et al., 2020; Xue et al., 2021). Ubiquitination of histones can regulate the stability and activity of specific proteins involved in gene regulation (Ding et al., 2012; Parida and Das, 2005).

The impact of histone modifications on gene expression during heat stress is complex and context-dependent. Different genes may exhibit distinct responses to histone modifications, and the interplay between various histone modifications can further modulate gene expression patterns (Ding et al., 2009; Kim et al., 2019). The "histone code" hypothesis also suggests that the combinatorial presence of multiple histone modifications at specific genomic regions can convey specific transcriptional states (Nabi et al., 2019). Understanding the precise mechanisms by which histone modifications influence gene expression during heat stress requires comprehensive studies and detailed investigations of particular genes and their regulatory elements (Rai et al., 2020a; Yang et al., 2020). However, histone modifications provide a vital layer of epigenetic regulation that allows plants to respond dynamically to heat stress by activating stress-responsive genes and fine-tuning gene expression patterns (Lämke et al., 2016; Chen et al., 2012). Overall, the impact of histone modifications on gene expression during heat stress highlights the significance of epigenetic mechanisms in plant responses to environmental challenges. Deciphering the specific roles of different histone modifications and their interplay will provide valuable insights into the molecular basis of heat stress responses and potentially offer avenues for enhancing stress tolerance in crops through targeted manipulation of histone modifications.

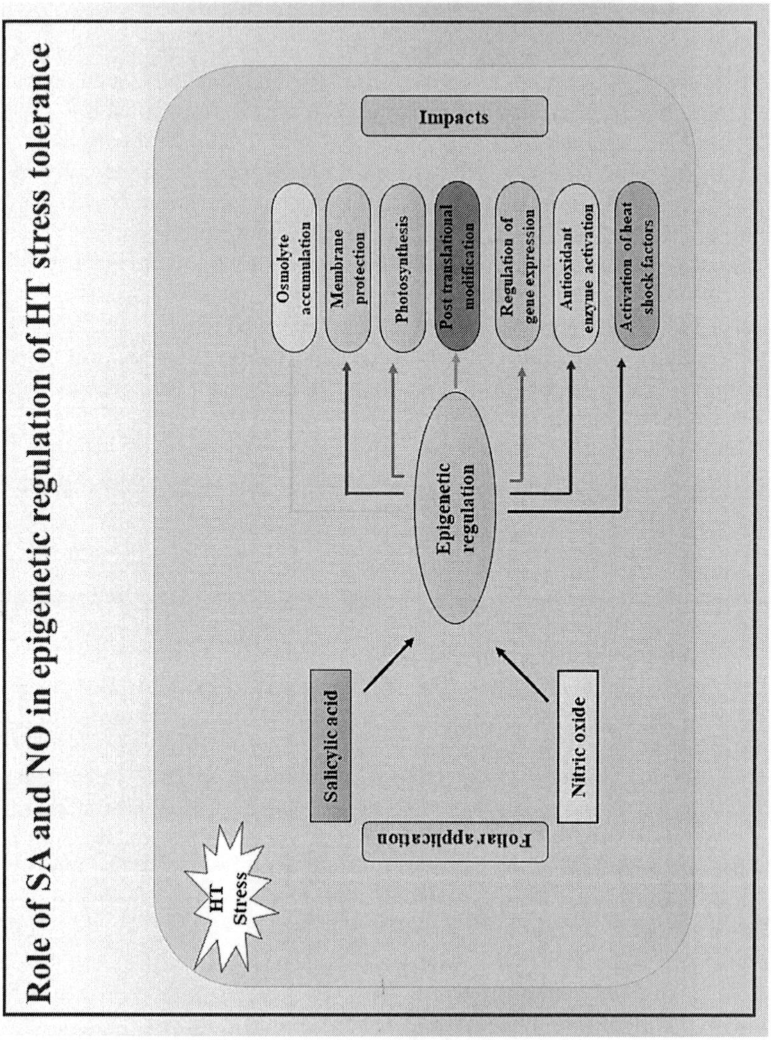


**Figure 1** Effect of salicylic acid and nitric oxide on modulation of epigentic regulation of high temperature stress tolerance in plants.

## Involvement of small RNAs in epigenetic regulation under heat stress

Under heat stress conditions, plants undergo various physiological and molecular changes to cope with the unfavorable environment. One key mechanism in regulating gene expression in response to heat stress is epigenetic regulation. Epigenetics refers to heritable changes in gene expression patterns that occur without alterations in the DNA sequence. Small RNAs, a class of non-coding RNAs, have emerged as important players in epigenetic regulation under heat stress in plants. Small RNAs, including microRNAs (miRNAs) and small interfering RNAs (siRNAs), are short RNA molecules that function as post-transcriptional regulators of gene expression. They are involved in diverse biological processes, including stress responses. Several studies have demonstrated that heat stress triggers changes in plants' abundance and activity of small RNAs (Bokszczanin et al., 2013; Yang et al., 2020; Mallory & Vaucheret, 2006; Sunkar et al., 2007; Xie et al., 2014).

In addition, plants produce specific small RNAs that can target and regulate the expression of genes involved in stress response pathways. These small RNAs can directly target the messenger RNAs (mRNAs) of specific genes for degradation or inhibit their translation into proteins. By modulating the expression of target genes, small RNAs can influence various physiological processes, such as heat tolerance, cell signaling, and metabolic adjustments (Ding et al., 2009; Mallory & Bouché, 2008; Sunkar et al., 2007).

Furthermore, small RNAs can also participate in epigenetic modifications by guiding DNA methylation and histone modifications. DNA methylation is an epigenetic modification that involves adding a methyl group to DNA, leading to changes in gene expression. Small RNAs can guide DNA methyltransferases to specific genomic loci, altering DNA methylation patterns (Henderson & Jacobsen, 2007; Rai et al., 2020). Similarly, small RNAs can interact with proteins involved in histone modifications, influencing DNA packaging into chromatin and affecting gene expression (Bies-Ethève et al., 2008; Li et al., 2013). Therefore, understanding the involvement of small RNAs in plant responses to heat stress can provide valuable insights into the molecular mechanisms underlying stress adaptation and potentially lead to the development of strategies for enhancing heat tolerance in crops.

## Conclusion

Plants have evolved sophisticated defense mechanisms, including stress memory, to cope with environmental stresses. The signaling molecules SA and NO critically affect stress memory development and heat stress responses. SA activates defense genes, enhances antimicrobial compound production, and primes plants for future stress encounters. NO mediates local and systemic defense responses and contributes to transgenerational heat stress memory by modulating epigenetic modifications. The interplay between SA and NO involves their production during stress, their interaction in regulating stress-responsive genes, and their synergistic effects on defence mechanisms. Understanding the complex signaling network and molecular mechanisms underlying the interplay between SA, NO, and epigenetic modifications is essential for improving crop resilience to environmental challenges. Manipulating these signaling pathways and epigenetic mechanisms may offer strategies to enhance plant stress tolerance. Overall, studying the role of SA and NO in stress memory provides valuable insights into plant adaptation and opens avenues for developing innovative approaches to enhance crop productivity and sustainability in changing climates.

## References

Arasimowicz, M. and Floryszak-Wieczorek, J., 2007. Nitric oxide as a bioactive signaling molecule in plant stress responses. Plant science, 172(5), pp.876-887.

Asgher, M., Per, T.S., Masood, A., Fatma, M., Freschi, L., Corpas, F.J. and Khan, N.A., 2017. Nitric oxide signaling and its crosstalk with other plant growth regulators in plant responses to abiotic stress. Environmental Science and Pollution Research, 24, pp.2273-2285.

Avramova, Z., 2015. Transcriptional 'memory'of a stress: transient chromatin and memory (epigenetic) marks at stress-response genes. The Plant Journal, 83(1), pp.149-159.

Bai, S., Li, M., Yao, T., Wang, H., Zhang, Y., Xiao, L., Wang, J., Zhang, Z., Hu, Y., Liu, W. and He, Y., 2012. Nitric oxide restrain root growth by DNA damage induced cell cycle arrest in Arabidopsis thaliana. Nitric Oxide, 26(1), pp.54-60.

Bauer, N.C., Corbett, A.H. and Doetsch, P.W., 2015. The current state of eukaryotic DNA base damage and repair. Nucleic acids research, 43(21), pp.10083-10101.

Beckers, G.J.M. and Spoel, S.H., 2005. Fine-tuning plant defence signaling: salicylate versus jasmonate. Plant biology, pp.1-10.

Beligni, M.V. and Lamattina, L., 2001. Nitric oxide in plants: the history is just beginning. Plant, Cell & Environment, 24(3), pp.267-278.

Berr, A., Shafiq, S. and Shen, W.H., 2011. Histone modifications in transcriptional activation during plant development. Biochimica et Biophysica Acta (BBA)-Gene Regulatory Mechanisms, 1809(10), pp.567-576.

Bhattacharjee, S., 2012. The language of reactive oxygen species signaling in plants. Journal of Botany, 2012.

Bies Etheve, N., Gaubier-Comella, P., Debures, A., Lasserre, E., Jobet, E., Raynal, M., Cooke, R. and Delseny, M., 2008. Inventory, evolution and expression profiling diversity of the LEA (late embryogenesis abundant) protein gene family in Arabidopsis thaliana. Plant molecular biology, 67, pp.107-124.

Bita, C.E. and Gerats, T., 2013. Plant tolerance to high temperature in a changing environment: scientific fundamentals and production of heat stress-tolerant crops. Frontiers in plant science, 4, p.273.

Bokszczanin, K.L., Solanaceae Pollen Thermotolerance Initial Training Network (SPOT-ITN) Consortium and Fragkostefanakis, S., 2013. Perspectives on deciphering mechanisms underlying plant heat stress response and thermotolerance. Frontiers in Plant Science, 4, p.315.

Brzezinka, K., Altmann, S., Czesnick, H., Nicolas, P., Gorka, M., Benke, E., Kabelitz, T., Jähne, F., Graf, A., Kappel, C. and Bäurle, I., 2016. Arabidopsis FORGETTER1 mediates stress-induced chromatin memory through nucleosome remodeling. elife, 5, p.e17061.

Byrne, M.E., 2012. Making leaves. Current Opinion in Plant Biology, 15(1), pp.24-30.

Chen, H., Chu, P., Zhou, Y., Li, Y., Liu, J., Ding, Y., Tsang, E.W., Jiang, L., Wu, K. and Huang, S., 2012. Overexpression of AtOGG1, a DNA glycosylase/AP lyase, enhances seed longevity and abiotic stress tolerance in Arabidopsis. Journal of experimental botany, 63(11), pp.4107-4121.

Chen, X., Ding, A.B. and Zhong, X., 2020. Functions and mechanisms of plant histone deacetylases. Science China Life Sciences, 63, pp.206-216.

Chen, X., Lu, L., Mayer, K.S., and Scalf, M. et al. (2018). Dynamics of histone H3.3 deposition in Arabidopsis thaliana reveals epigenetic dynamics during plant reproduction. Nucleic Acids Res. 46, 432–445.

Chen, Z., 2021. Deciphering the impacts of heat stress on the acquisition of seed quality traits in *Medicago truncatula* (Doctoral dissertation, Agrocampus Ouest).

Chinnusamy, V. and Zhu, J.K., 2009. Epigenetic regulation of stress responses in plants. Current opinion in plant biology, 12(2), pp.133-139.

Cong, W., Miao, Y., Xu, L., Zhang, Y., Yuan, C., Wang, J., Zhuang, T., Lin, X., Jiang, L., Wang, N. and Ma, J., 2019. Transgenerational memory of gene expression changes induced by heavy metal stress in rice (*Oryza sativa* L.). BMC plant biology, 19(1), pp.1-14.

Corpas, F.J. and Barroso, J.B., 2017. Nitric oxide synthase-like activity in higher plants. Nitric oxide: biology and chemistry, 68, pp.5-6.

Corpas, F.J., Barroso, J.B., Palma, J.M. and Rodriguez-Ruiz, M., 2017. Plant peroxisomes: a nitro-oxidative cocktail. Redox Biol 11: 535–542.

Correa-Aragunde, N., Graziano, M. and Lamattina, L., 2004. Nitric oxide plays a central role in determining lateral root development in tomato. Planta, 218, pp.900-905.

Dempsey, D.M.A., Vlot, A.C., Wildermuth, M.C. and Klessig, D.F., 2011. Salicylic acid biosynthesis and metabolism. The Arabidopsis book/American Society of Plant Biologists, 9.

Ding, D., Zhang, L., Wang, H., Liu, Z., Zhang, Z. and Zheng, Y., 2009. Differential expression of miRNAs in response to salt stress in maize roots. Annals of botany, 103(1), pp.29-38.

Ding, Y., Fromm, M. and Avramova, Z., 2012. Multiple exposures to drought'train'transcriptional responses in Arabidopsis. Nature communications, 3(1), p.740.

Ding, Y., Liu, N., Virlouvet, L., Riethoven, J.J., Fromm, M. and Avramova, Z., 2013. Four distinct types of dehydration stress memory genes in Arabidopsis thaliana. BMC plant biology, 13, pp.1-11.

Dowen, R.H., Pelizzola, M., Schmitz, R.J., Lister, R., Dowen, J.M., Nery, J.R., Dixon, J.E. and Ecker, J.R., 2012. Widespread dynamic DNA methylation in response to biotic stress. Proceedings of the National Academy of Sciences, 109(32), pp. E2183-E2191.

Dowen, R.H., Pelizzola, M., Schmitz, R.J., Lister, R., Dowen, J.M., Nery, J.R., Dixon, J.E. and Ecker, J.R., 2012. Widespread dynamic DNA methylation in response to biotic stress. Proceedings of the National Academy of Sciences, 109(32), pp.E2183-E2191.

Engler, D. and Jakob, K., 2013. Genetic engineering of Miscanthus. Genomics of the Saccharinae, pp.255-301.

Essemine, J., Ammar, S. and Bouzid, S., 2010. Impact of heat stress on germination and growth in higher plants: Physiological, biochemical and molecular repercussions and mechanisms of defence. Journal of Biological Sciences, 10(6), pp.565-572.

Feng, B., Zhang, C., Chen, T., Zhang, X., Tao, L. and Fu, G., 2018. Salicylic acid reverses pollen abortion of rice caused by heat stress. BMC Plant Biology, 18(1), pp.1-16.

Feng, S., Cokus, S.J., Zhang, X., Chen, P.Y., Bostick, M., Goll, M.G., Hetzel, J., Jain, J., Strauss, S.H., Halpern, M.E. and Ukomadu, C., 2010. Conservation and divergence of methylation patterning in plants and animals. Proceedings of the National Academy of Sciences, 107(19), pp.8689-8694.

Fragkostefanakis, S., Mesihovic, A., Simm, S., Paupière, M.J., Hu, Y., Paul, P., Mishra, S.K., Tschiersch, B., Theres, K., Bovy, A. and Schleiff, E., 2016. HsfA2 controls the activity of developmentally and stress-regulated heat stress protection mechanisms in tomato male reproductive tissues. Plant physiology, 170(4), pp.2461-2477.

Fritsche, S., Wang, X. and Jung, C., 2017. Recent advances in our understanding of tocopherol biosynthesis in plants: an overview of key genes, functions, and breeding of vitamin E improved crops. Antioxidants, 6(4), p.99.

Gao, Z., Zhou, Y. and He, Y., 2022. Molecular epigenetic mechanisms for memory of temperature stresses in plants. Journal of Genetics and Genomics.

Garg, R., Narayana Chevala, V.V.S., Shankar, R. and Jain, M., 2015. Divergent DNA methylation patterns associated with gene expression in rice cultivars with contrasting drought and salinity stress response. Scientific reports, 5(1), p.14922.

Grima-Pettenati, J., Soler, M., Camargo, E.L.O. and Wang, H., 2012. Transcriptional regulation of the lignin biosynthetic pathway revisited: new players and insights. In Advances in botanical research (Vol. 61, pp. 173-218). Academic Press.

Hauben, M., Haesendonckx, B., Standaert, E., Van Der Kelen, K., Azmi, A., Akpo, H., Van Breusegem, F., Guisez, Y., Bots, M., Lambert, B. and Laga, B., 2009. Energy use efficiency is characterized by an epigenetic component that can be directed through artificial selection to increase yield. Proceedings of the National Academy of Sciences, 106(47), pp.20109-20114.

Hauser, M.T., Aufsatz, W., Jonak, C. and Luschnig, C., 2011. Transgenerational epigenetic inheritance in plants. Biochimica et Biophysica Acta (BBA)-Gene Regulatory Mechanisms, 1809(8), pp.459-468.

He, S., Zhang, Y., Wang, J., Wang, Y., Ji, F., Sun, L., Zhang, G. and Hao, F., 2022. H3K4me2, H4K5ac and DNA methylation function in short- and long-term heat stress responses through affecting the expression of the stress-related genes in *G. hirsutum*. Environmental and Experimental Botany, 194, p.104699.

Henderson, I.R. and Jacobsen, S.E., 2007. Epigenetic inheritance in plants. Nature, 447(7143), pp.418-424.

Hilker, M., Schwachtje, J., Baier, M., Balazadeh, S., Bäurle, I., Geiselhardt, S., Hincha, D.K., Kunze, R., Mueller-Roeber, B., Rillig, M.C. and Rolff, J., 2016. Priming and memory of stress responses in organisms lacking a nervous system. Biological Reviews, 91(4), pp.1118-1133.

Hu, J., Yang, H., Mu, J., Lu, T., Peng, J., Deng, X., Kong, Z., Bao, S., Cao, X. and Zuo, J., 2017. Nitric oxide regulates protein methylation during stress responses in plants. Molecular Cell, 67(4), pp.702-710.

Hu, L., Li, N., Xu, C., Zhong, S., Lin, X., Yang, J., Zhou, T., Yuliang, A., Wu, Y., Chen, Y.R. and Cao, X., 2014. Mutation of a major CG methylase in rice causes genome-wide hypomethylation, dysregulated genome expression, and seedling lethality. Proceedings of the National Academy of Sciences, 111(29), pp.10642-10647.

Ito, H., Gaubert, H., Bucher, E., Mirouze, M., Vaillant, I. and Paszkowski, J., 2011. An siRNA pathway prevents transgenerational retrotransposition in plants subjected to stress. Nature, 472(7341), pp.115-119.

Iwasaki, M. and Paszkowski, J., 2014. Epigenetic memory in plants. The EMBO journal, 33(18), pp.1987-1998.

Iwasaki, M. and Paszkowski, J., 2014. Epigenetic memory in plants. The EMBO journal, 33(18), pp.1987-1998.

Jaskiewicz, M., Conrath, U. and Peterhänsel, C., 2011. Chromatin modification acts as a memory for systemic acquired resistance in the plant stress response. EMBO reports, 12(1), pp.50-55.

Johannes, F., Porcher, E., Teixeira, F.K., Saliba-Colombani, V., Simon, M., Agier, N., Bulski, A., Albuisson, J., Heredia, F., Audigier, P. and Bouchez, D., 2009. Assessing the impact of transgenerational epigenetic variation on complex traits. PLoS genetics, 5(6), p.e1000530.

Khan, M.I.R., Fatma, M., Per, T.S., Anjum, N.A. and Khan, N.A., 2015. Salicylic acid-induced abiotic stress tolerance and underlying mechanisms in plants. Frontiers in plant science, 6, p.462.

Khanna, K., Bhardwaj, R., Alam, P., Reiter, R.J. and Ahmad, P., 2023. Phytomelatonin: A master regulator for plant oxidative stress management. Plant Physiology and Biochemistry.

KIM, J.M., To, T.K., Nishioka, T. and Seki, M., 2010. Chromatin regulation functions in plant abiotic stress responses. Plant, Cell & Environment, 33(4), pp.604-611.

Kinoshita, T. and Seki, M., 2014. Epigenetic memory for stress response and adaptation in plants. Plant and Cell Physiology, 55(11), pp.1859-1863.

Klessig, D.F., Choi, H.W. and Dempsey, D.M.A., 2018. Systemic acquired resistance and salicylic acid: past, present, and future. Molecular plant-microbe interactions, 31(9), pp.871-888.

Klessig, D.F., Durner, J., Shah, J. and Yang, Y., 1998. Salicylic acid-mediated signal transduction in plant disease resistance. Phytochemical signals and plant-microbe interactions, pp.119-137.

Kohli, S.K., Khanna, K., Bhardwaj, R., Corpas, F.J. and Ahmad, P., 2022. Nitric oxide, salicylic acid and oxidative stress: Is it a perfect equilateral triangle?. Plant Physiology and Biochemistry, 184, pp.56-64.

Kooke, R., Johannes, F., Wardenaar, R., Becker, F., Etcheverry, M., Colot, V., Vreugdenhil, D. and Keurentjes, J.J., 2015. Epigenetic basis of morphological variation and phenotypic plasticity in Arabidopsis thaliana. The Plant Cell, 27(2), pp.337-348.

Korotko, U., Chwiałkowska, K., Sańko-Sawczenko, I. and Kwasniewski, M., 2021. DNA demethylation in response to heat stress in Arabidopsis thaliana. International journal of molecular sciences, 22(4), p.1555.

Kovacs, I., Durner, J. and Lindermayr, C., 2015. Crosstalk between nitric oxide and glutathione is required for NONEXPRESSOR OF PATHOGENESIS-RELATED GENES 1 (NPR 1)-dependent defense signaling in Arabidopsis thaliana. New Phytologist, 208(3), pp.860-872.

Lämke, J. and Bäurle, I., 2017. Epigenetic and chromatin-based mechanisms in environmental stress adaptation and stress memory in plants. Genome biology, 18(1), pp.1-11.

Lang-Mladek, C., Popova, O., Kiok, K., Berlinger, M., Rakic, B., Aufsatz, W., Jonak, C., Hauser, M.T. and Luschnig, C., 2010. Transgenerational inheritance and resetting of stress-induced loss of epigenetic gene silencing in Arabidopsis. Molecular plant, 3(3), pp.594-602.

Latzel, V.I.T., Allan, E., Bortolini Silveira, A., Colot, V., Fischer, M. and Bossdorf, O., 2013. Epigenetic diversity increases the productivity and stability of plant populations. Nature communications, 4(1), p.2875.

Law, J.A. and Jacobsen, S.E., 2010. Establishing, maintaining and modifying DNA methylation patterns in plants and animals. Nature Reviews Genetics, 11(3), pp.204-220.

Lee, S., Lee, H.J., Huh, S.U., Paek, K.H., Ha, J.H. and Park, C.M., 2014. The Arabidopsis NAC transcription factor NTL4 participates in a positive feedback loop that induces programmed cell death under heat stress conditions. Plant Science, 227, pp.76-83.

Lee, S., Lee, H.J., Huh, S.U., Paek, K.H., Ha, J.H. and Park, C.M., 2014. The Arabidopsis NAC transcription factor NTL4 participates in a positive feedback loop that induces programmed cell death under heat stress conditions. Plant Science, 227, pp.76-83.

Li, N., Han, X., Feng, D., Yuan, D. and Huang, L.J., 2019. Signaling crosstalk between salicylic acid and ethylene/jasmonate in plant defense: do we understand what they are whispering?. International journal of molecular sciences, 20(3), p.671.

Li, S., Liu, L., Zhuang, X., Yu, Y., Liu, X., Cui, X., Ji, L., Pan, Z., Cao, X., Mo, B. and Zhang, F., 2013. MicroRNAs inhibit the translation of target mRNAs on the endoplasmic reticulum in Arabidopsis. Cell, 153(3), pp.562-574.

Liu, H., Able, A.J. and Able, J.A., 2021. Small RNA, transcriptome and degradome analysis of the transgenerational heat stress response network in durum wheat. International Journal of Molecular Sciences, 22(11), p.5532.

Liu, P.P., Montgomery, T.A., Fahlgren, N., Kasschau, K.D., Nonogaki, H. and Carrington, J.C., 2007. Repression of AUXIN RESPONSE FACTOR10 by microRNA160 is critical for seed germination and post-germination stages. The Plant Journal, 52(1), pp.133-146.

Liu, R. and Lang, Z., 2020. The mechanism and function of active DNA demethylation in plants. Journal of integrative plant biology, 62(1), pp.148-159.

Liu, X., Dong, J., Zhang, L., et al. (2019). Transgenerational memory of gene expression changes induced by heavy metal stress in rice (Oryza sativa L.). BMC Plant Biology, 19(1), 477. doi: 10.1186/s12870-019-2074-9

Liu, X., Zhang, W., Zhao, M., et al. (2020b). Nitric oxide regulates DNA methylation during heat stress in Arabidopsis. Plant Physiology, 183(1), 245-258. doi: 10.1104/pp.19.01443

Lu, Y., Bu, Q., Chuan, M., Cui, X., Zhao, Y. and Zhou, D.X., 2023. Metabolic regulation of the plant epigenome. The Plant Journal, 114(5), pp.1001-1013.

Mallory, A.C. and Bouché, N., 2008. MicroRNA-directed regulation: to cleave or not to cleave. Trends in plant science, 13(7), pp.359-367.

Mallory, A.C. and Vaucheret, H., 2006. Functions of microRNAs and related small RNAs in plants. Nature genetics, 38(Suppl 6), pp.S31-S36.

Mammarella, M.F., Lucero, L., Hussain, N., Muñoz-Lopez, A., Huang, Y., Ferrero, L., Fernandez-Milmanda, G., Manavella, P.A., Benhamed, M., Crespi, M. and Ballare, C., 2023. Long noncoding RNA-mediated epigenetic regulation of auxin-related genes controlling shade avoidance syndrome in *Arabidopsis thaliana*. bioRxiv, pp.2023-03.

Martinez-Medina, A., Flors, V., Heil, M., Mauch-Mani, B., Pieterse, C.M., Pozo, M.J., Ton, J., van Dam, N.M. and Conrath, U., 2016. Recognizing plant defense priming. Trends in plant science, 21(10), pp.818-822.

Matzke, M., Kanno, T., Daxinger, L., Huettel, B. and Matzke, A.J., 2009. RNA-mediated chromatin-based silencing in plants. Current opinion in cell biology, 21(3), pp.367-376.

Mengel, A., Ageeva, A., Georgii, E., Bernhardt, J., Wu, K., Durner, J. and Lindermayr, C., 2017. Nitric oxide modulates histone acetylation at stress genes by inhibition of histone deacetylases. Plant Physiology, 173(2), pp.1434-1452.

Mirouze, M. and Paszkowski, J., 2011. Epigenetic contribution to stress adaptation in plants. Current opinion in plant biology, 14(3), pp.267-274.

Miura, K. and Tada, Y., 2014. Regulation of water, salinity, and cold stress responses by salicylic acid. Frontiers in plant science, 5, p.4.

Munir, M.U.B.A.S.H.R.A.H. and Shabbir, G.H.U.L.A.M., 2018. Salicylic acid mediated heat stress tolerance in selected bread wheat genotypes of Pakistan. Pak. J. Bot, 50(6), pp.2141-2146.

Nabi, R.B.S., Tayade, R., Hussain, A., Kulkarni, K.P., Imran, Q.M., Mun, B.G. and Yun, B.W., 2019. Nitric oxide regulates plant responses to drought, salinity, and heavy metal stress. Environmental and Experimental Botany, 161, pp.120-133.

Nabi, R.B.S., Tayade, R., Hussain, A., Kulkarni, K.P., Imran, Q.M., Mun, B.G. and Yun, B.W., 2019. Nitric oxide regulates plant responses to drought, salinity, and heavy metal stress. Environmental and Experimental Botany, 161, pp.120-133.

Neill, S.J., Desikan, R., Clarke, A. and Hancock, J.T., 2002. Nitric oxide is a novel component of abscisic acid signaling in stomatal guard cells. Plant physiology, 128(1), pp.13-16.

Otero, S., Desvoyes, B. and Gutierrez, C., 2014. Histone H3 dynamics in plant cell cycle and development. Cytogenetic and genome research, 143(1-3), pp.114-124.

Paluvai, H., Di Giorgio, E. and Brancolini, C., 2020. The histone code of senescence. Cells, 9(2), p.466.

Parankusam, S., Adimulam, S.S., Bhatnagar-Mathur, P. and Sharma, K.K., 2017. Nitric oxide (NO) in plant heat stress tolerance: current knowledge and perspectives. Frontiers in Plant Science, 8, p.1582.

Parida, A.K. and Das, A.B., 2005. Salt tolerance and salinity effects on plants: a review. Ecotoxicology and environmental safety, 60(3), pp.324-349.

Pérez-Salamó, I., Papdi, C., Rigó, G., Zsigmond, L., Vilela, B., Lumbreras, V., Nagy, I., Horváth, B., Domoki, M., Darula, Z. and Medzihradszky, K., 2014. The heat shock factor A4A confers salt tolerance and is regulated by oxidative stress and the mitogen-activated protein kinases MPK3 and MPK6. Plant physiology, 165(1), pp.319-334.

Pfluger, J. and Wagner, D., 2007. Histone modifications and dynamic regulation of genome accessibility in plants. Current opinion in plant biology, 10(6), pp.645-652.

Pick, T., Jaskiewicz, M., Peterhänsel, C. and Conrath, U., 2012. Heat shock factor HsfB1 primes gene transcription and systemic acquired resistance in Arabidopsis. Plant Physiology, 159(1), pp.52-55.

Prakash, V., Singh, V.P., Tripathi, D.K., Sharma, S. and Corpas, F.J., 2021. Nitric oxide (NO) and salicylic acid (SA): A framework for their relationship in plant development under abiotic stress. Plant Biology, 23, pp.39-49.

Qiao, W. and Fan, L.M., 2008. Nitric oxide signaling in plant responses to abiotic stresses. Journal of Integrative Plant Biology, 50(10), pp.1238-1246.

Quadrana, L. and Colot, V., 2016. Plant transgenerational epigenetics. Annual Review of Genetics, 50, pp.467-491.

Queitsch, C., Hong, S.W., Vierling, E. and Lindquist, S., 2000. Heat shock protein 101 plays a crucial role in thermotolerance in Arabidopsis. The Plant Cell, 12(4), pp.479-492.

Rai, K.K., Pandey, N. and Rai, S.P., 2020. Salicylic acid and nitric oxide signaling in plant heat stress. Physiologia plantarum, 168(2), pp.241-255.

Rai, K.K., Pandey, N., Rai, N., Rai, S.K. and Pandey-Rai, S., 2021. Salicylic acid and nitric oxide: insight into the transcriptional regulation of their metabolism and regulatory functions in plants. Frontiers in Agronomy, 3, p.781027.

Rai, K.K., Rai, N. and Rai, S.P., 2018. Salicylic acid and nitric oxide alleviate high temperature induced oxidative damage in *Lablab purpureus* L plants by regulating bio-physical processes and DNA methylation. Plant Physiology and Biochemistry, 128, pp.72-88.

Rai, K.K., Rai, N. and Rai, S.P., 2018. Salicylic acid and nitric oxide alleviate high temperature induced oxidative damage in Lablab purpureus L plants by regulating bio-physical processes and DNA methylation. Plant Physiology and Biochemistry, 128, pp.72-88.

Rai, K.K., Rai, N., Aamir, M., Tripathi, D. and Rai, S.P., 2020a. Interactive role of salicylic acid and nitric oxide on transcriptional reprogramming for high temperature tolerance in *Lablab purpureus* L.: Structural and functional insights using computational approaches. Journal of biotechnology, 309, pp.113-130.

Rivas-San Vicente, M. and Plasencia, J., 2011. Salicylic acid beyond defence: its role in plant growth and development. Journal of experimental botany, 62(10), pp.3321-3338.

Rudolf, E.E., Hüther, P., Forné, I., Georgii, E., Han, Y., Hell, R., Wirtz, M., Imhof, A., Becker, C., Durner, J. and Lindermayr, C., 2021. GSNOR contributes to demethylation and expression of transposable elements and stress-responsive genes. Antioxidants, 10(7), p.1128.

Růžička, K., Zhang, M., Campilho, A., Bodi, Z., Kashif, M., Saleh, M., Eeckhout, D., El-Showk, S., Li, H., Zhong, S. and De Jaeger, G., 2017. Identification of factors required for m6A mRNA methylation in Arabidopsis reveals a role for the conserved E3 ubiquitin ligase HAKAI. New Phytologist, 215(1), pp.157-172.

Sang, J., Jiang, M., Lin, F., Xu, S., Zhang, A. and Tan, M., 2008. Nitric oxide reduces hydrogen peroxide accumulation involved in water stress-induced subcellular anti-oxidant defense in maize plants. Journal of Integrative Plant Biology, 50(2), pp.231-243.

Secco, D., Jabnoune, M., Walker, H., Shou, H., Wu, P., Poirier, Y. and Whelan, J., 2013. Spatio-temporal transcript profiling of rice roots and shoots in response to phosphate starvation and recovery. The Plant Cell, 25(11), pp.4285-4304.

Seo, P.J., Kim, M.J., Park, J.Y., Kim, S.Y., Jeon, J., Lee, Y.H., Kim, J. and Park, C.M., 2010. Cold activation of a plasma membrane-tethered NAC transcription factor induces a pathogen resistance response in Arabidopsis. The Plant Journal, 61(4), pp.661-671.

Sequeira-Mendes, J., Aragüez, I., Peiró, R., Mendez-Giraldez, R., Zhang, X., Jacobsen, S.E., Bastolla, U. and Gutierrez, C., 2014. The functional topography of the Arabidopsis genome is organized in a reduced number of linear motifs of chromatin states. The Plant Cell, 26(6), pp.2351-2366.

Shanker, A.K., Bhanu, D. and Maheswari, M., 2020. Epigenetics and transgenerational memory in plants under heat stress. Plant Physiology Reports, 25(4), pp.583-593.

Shapiro, A.D., 2005. Nitric oxide signaling in plants. Vitamins & Hormones, 72, pp.339-398.

Sharma, M., Sharma, M., Jamsheer K, M. and Laxmi, A., 2022. A glucose–target of rapamycin signaling axis integrates environmental history of heat stress through maintenance of transcription-associated epigenetic memory in Arabidopsis. Journal of Experimental Botany, 73(20), pp.7083-7102.

Shekhawat, K., Almeida-Trapp, M., García-Ramírez, G.X. and Hirt, H., 2022. Beat the heat: Plant-and microbe-mediated strategies for crop thermotolerance. Trends in Plant Science, 27(8), pp.802-813.

Shu, K., Qi, Y., Chen, F., Meng, Y., Luo, X., Shuai, H., Zhou, W., Ding, J., Du, J., Liu, J. and Yang, F., 2017. Salt stress represses soybean seed germination by negatively regulating GA biosynthesis while positively mediating ABA biosynthesis. Frontiers in plant science, 8, p.1372.

Singroha, G. and Sharma, P., 2019. Epigenetic modifications in plants under abiotic stress. Epigenetics, 10.

Slaughter, A., Daniel, X., Flors, V., Luna, E., Hohn, B. and Mauch-Mani, B., 2012. Descendants of primed Arabidopsis plants exhibit resistance to biotic stress. Plant physiology, 158(2), pp.835-843.

Song, L., Ding, W., Shen, J., Zhang, Z., Bi, Y. and Zhang, L., 2008. Nitric oxide mediates abscisic acid induced thermotolerance in the calluses from two ecotypes of reed under heat stress. Plant Science, 175(6), pp.826-832.

Song, Y., Ji, D., Li, S., Wang, P., Li, Q. and Xiang, F., 2012. The dynamic changes of DNA methylation and histone modifications of salt responsive transcription factor genes in soybean. PloS one, 7(7), p.e41274.

Springer, N.M., 2013. Epigenetics and crop improvement. Trends in genetics, 29(4), pp.241-247.

Sunkar, R., Chinnusamy, V., Zhu, J. and Zhu, J.K., 2007. Small RNAs as big players in plant abiotic stress responses and nutrient deprivation. Trends in plant science, 12(7), pp.301-309.

Vlachonasios, K. E., Thomashow, M. F., & Triezenberg, S. J. (2011). Disruption mutants of Arabidopsis MORC1 reveal the role of MORC1 in DNA methylation. The Plant Cell, 23(3), 1010-1028. doi: 10.1105/tpc.111.083915

Vlot, A.C., Dempsey, D.M.A. and Klessig, D.F., 2009. Salicylic acid, a multifaceted hormone to combat disease. Annual review of phytopathology, 47, pp.177-206.

Wahid, A., Gelani, S., Ashraf, M. and Foolad, M.R., 2007. Heat tolerance in plants: an overview. Environmental and experimental botany, 61(3), pp.199-223.

Wang, L.J. and Li, S.H., 2006. Thermotolerance and related antioxidant enzyme activities induced by heat acclimation and salicylic acid in grape (Vitis vinifera L.) leaves. Plant Growth Regulation, 48, pp.137-144.

Wang, P., Du, Y., Hou, Y.J., Zhao, Y., Hsu, C.C., Yuan, F., Zhu, X., Tao, W.A., Song, C.P. and Zhu, J.K., 2015. Nitric oxide negatively regulates abscisic acid signaling in guard cells by S-nitrosylation of OST1. Proceedings of the National Academy of Sciences, 112(2), pp.613-618.

Wendehenne, D., Gao, Q.M., Kachroo, A. and Kachroo, P., 2014. Free radical-mediated systemic immunity in plants. Current opinion in plant biology, 20, pp.127-134.

Xie, F., Stewart Jr, C. N., Taki, F. A., & He, Q. (2014). Silencing of root-knot nematode genes by siRNA in transformed soybean roots. Experimental Parasitology, 145, 38-42.
doi: 10.1016/j.exppara.2014.06.002

Xue, Y., Zhong, Z., Harris, C.J., Gallego-Bartolomé, J., Wang, M., Picard, C., Cao, X., Hua, S., Kwok, I., Feng, S. and Jami-Alahmadi, Y., 2021. Arabidopsis MORC proteins function in the efficient establishment of RNA directed DNA methylation. Nature communications, 12(1), p.4292.

Yang, J., Yuan, L., Yen, M.R., Zheng, F., Ji, R., Peng, T., Gu, D., Yang, S., Cui, Y., Chen, P.Y. and Wu, K., 2020. SWI3B and HDA6 interact and are required for transposon silencing in Arabidopsis. The Plant Journal, 102(4), pp.809-822.

Yang, X., Zhang, L., Yang, Y., Schmid, M. and Wang, Y., 2021. miRNA mediated regulation and interaction between plants and pathogens. International Journal of Molecular Sciences, 22(6), p.2913.

Yong-Villalobos, L., González-Morales, S.I., Wrobel, K., Gutiérrez-Alanis, D., Cervantes-Peréz, S.A., Hayano-Kanashiro, C., Oropeza-Aburto, A., Cruz-Ramírez, A., Martínez, O. and Herrera-Estrella, L., 2015. Methylome analysis reveals an important role for epigenetic changes in the regulation of the Arabidopsis response to phosphate starvation. Proceedings of the National Academy of Sciences, 112(52), pp.E7293-E7302.

Yuan, L., Liu, X., Luo, M., Yang, S. and Wu, K., 2013. Involvement of histone modifications in plant abiotic stress responses. Journal of Integrative Plant Biology, 55(10), pp.892-901.

Yun, B.W., Feechan, A., Yin, M., Saidi, N.B., Le Bihan, T., Yu, M., Moore, J.W., Kang, J.G., Kwon, E., Spoel, S.H. and Pallas, J.A., 2011. S-nitrosylation of NADPH oxidase regulates cell death in plant immunity. Nature, 478(7368), pp.264-268.

Zhang, X., Clarenz, O., Cokus, S., Bernatavichute, Y.V., Pellegrini, M., Goodrich, J. and Jacobsen, S.E., 2007. Whole-genome analysis of histone H3 lysine 27 trimethylation in Arabidopsis. PLoS biology, 5(5), p.e129.

Zhou, M., Wang, W., Karapetyan, S., Mwimba, M., Marqués, J., Buchler, N.E. and Dong, X., 2015. Redox rhythm reinforces the circadian clock to gate immune response. Nature, 523(7561), pp.472-476.

Zhou, Y., Chen, H., Chu, P., Li, Y., Tan, B., Ding, Y., Tsang, E.W., Jiang, L., Wu, K. and Huang, S., 2012. NnHSP17. 5, a cytosolic class II small heat shock protein gene from Nelumbo nucifera, contributes to seed germination vigor and seedling thermotolerance in transgenic Arabidopsis. Plant cell reports, 31, pp.379-389.

# CHAPTER 8

# SALICYLIC ACID AND NITRIC OXIDE: REGULATION OF SMALL RNAS AND TRANSPOSABLE ELEMENTS UNDER HEAT STRESS


## Abstract

Salicylic acid (SA) and nitric oxide (NO) are signaling molecules involved in regulating plant responses to heat stress. Small RNAs (sRNAs) and transposable elements (TEs) are key players in gene regulation and genome stability. The interplay between SA and NO signaling pathways influences the expression of sRNAs and the activity of TEs under heat stress conditions. SA accumulation under heat stress increases sRNA expression, suppressing TE activity and ensuring genome stability. Similarly, NO production increases under heat stress and enhances sRNA accumulation, leading to the silencing of TEs. The synergistic effect of SA and NO pathways contributes to the coordinated regulation of sRNAs and TEs, involving complex molecular interactions and transcriptional regulation mechanisms. This chapter explores the intricate regulatory networks of SA, NO, sRNAs, and TEs in enhancing plant stress tolerance and offers insights for improving crop productivity in challenging environments.



**Keywords**: Salicylic acid, Nitric oxide, Small RNAs, Transposable elements, Heat stress


## Introduction

SA and NO are two essential signaling molecules that regulate various physiological processes in plants, including responses to heat stress (Rai et al., 2020; Prakash et al., 2021). Heat stress is a significant environmental factor adversely affecting plant growth and development. Understanding how SA and NO regulate gene expression under heat stress conditions is crucial for improving plant resilience and stress tolerance. Small RNAs are short RNA molecules critical in gene regulation. They can target and silence specific genes or TEs through RNA interference. TEs are mobile genetic elements that can move within a genome and potentially disrupt gene expression. The regulation of sRNAs and TEs under heat stress involves the interplay of SA and NO signaling pathways (Liu et al., 2017).

SA accumulates in plants upon exposure to various stress conditions, including heat stress. It acts as a critical signaling molecule that regulates defense responses. SA has been found to modulate the expression of heat stress-responsive genes and sRNAs (Miura and Tada, 2014). It has been observed that SA accumulation under heat stress conditions increases the expression of sRNAs, suppressing the activity of TEs (Mao et al., 2015). This mechanism helps maintain genome stability and prevent potential damage caused by TEs. NO is another critical signaling molecule involved in diverse physiological processes, including stress responses. Under heat stress, NO production increases in plants, acting as a signaling molecule to induce heat stress tolerance (Shivaraj et al., 2020). NO can regulate the expression of heat stress-responsive genes and sRNAs (Prakash et al., 2021). Studies have shown that NO can enhance the accumulation of sRNAs, which target and silence TEs, reducing their potential adverse effects on gene expression under heat stress conditions (Huang et al., 2022).

The interplay between SA and NO signaling pathways is essential for the coordinated regulation of sRNAs and TEs under heat stress (Rai et al., 2020). SA and NO can influence each other's signaling pathways, leading to a synergistic effect on gene expression regulation (Meng et al., 2022). The cross-talk between SA and NO pathways involves complex molecular interactions and transcriptional regulatory mechanisms (Siddiqui et al., 2011). SA and NO are important in regulating sRNAs and TEs under heat stress conditions. They act as signaling molecules that modulate gene expression and contribute to plant stress tolerance. This chapter will unravel the intricate regulatory networks involving SA, NO, sRNAs, and TEs in enhancing plant stress tolerance. Further, it will also provide valuable insights for developing strategies to enhance plant resilience to heat stress and improve crop productivity in challenging environments.

## Small RNAs: Roles and Biogenesis

Small RNAs are short RNA molecules that play critical roles in various biological processes, including the response to heat stress in plants. They regulate genes by targeting specific mRNA molecules for degradation or translational inhibition (Li et al., 2010). Understanding the roles and biogenesis of sRNAs in the heat stress response is crucial for elucidating the molecular mechanisms underlying plant adaptation to high temperatures (Zhao et al., 2022). During heat stress, plants undergo significant physiological and biochemical changes to cope with adverse conditions. The production and accumulation of sRNAs are part of the intricate regulatory network that enables plants to modulate gene expression and enhance stress tolerance (Shi et al., 2018). Two major classes of sRNAs involved in the heat stress response are microRNAs (miRNAs) and small interfering RNAs (siRNAs) (Li et al., 2020).

MiRNAs are sRNAs that regulate gene expression at the post-transcriptional level. They are generated from longer primary transcripts that fold into hairpin structures, known as precursor miRNAs (pre-miRNAs). These pre-miRNAs are processed by an enzyme complex called DICER-LIKE 1 (DCL1) to produce mature miRNAs (Axtell and Bartel, 2005). The mature miRNAs then guide the RNA-induced silencing complex (RISC) to target specific mRNAs for degradation or translational repression (Khraiwesh et al., 2012). Several miRNAs have been identified to be involved in heat stress response, and their targets include genes encoding transcription factors, heat shock proteins, and other heat stress-related proteins (Yadav et al., 2015; Nadarajah and Kumar, 2019). SiRNAs, on the other hand, are derived from various sources, including repetitive DNA sequences, TEs, and viral RNA. They play a crucial role in genome defense by silencing TEs and maintaining genome stability (Zhai et al., 2015). During heat stress, the expression of TEs can be induced, posing a potential threat to genome integrity. The biogenesis of siRNAs involves the action of RNA-dependent RNA polymerases (RDRs), which generate double-stranded RNAs (dsRNAs) from TE transcripts. These dsRNAs are subsequently processed into siRNAs by DCL proteins (Zhao et al., 2022). The siRNAs then guide the RISC to target complementary sequences in TE transcripts, leading to their degradation or silencing (Sunkar et al., 2007).

Overall, sRNAs, including miRNAs and siRNAs, are integral components of the plant heat stress response. They regulate gene expression by targeting specific mRNAs or silencing TEs, modulating vital cellular processes and enhancing stress tolerance. Elucidating sRNAs' biogenesis and functional roles in heat stress response will provide valuable insights into the molecular mechanisms underlying plant adaptation to high temperatures. It may contribute to developing strategies for improving crop resilience to heat stress.

## Importance of small RNAs in post-transcriptional gene regulation and stress responses

Small RNAs are crucial in post-transcriptional gene regulation and stress responses in organisms (Borges and Martienssen, 2015). These short RNA molecules, typically ranging from 20 to 30 nucleotides in length, are involved in fine-tuning gene expression and maintaining cellular homeostasis under normal and stressful conditions (Carthew and Sontheimer, 2009). Post-transcriptional gene regulation refers to the processes that occur after transcription, where the genetic information encoded in DNA is transcribed into mRNA (Ha and Kim, 2014). Small RNAs, such as miRNAs and small siRNAs, are critical players in this regulatory mechanism (Axtell and Bowman, 2008). They bind to target mRNA molecules through sequence complementarity, leading to either mRNA degradation or translational repression (Bartel, 2009). The importance of small RNAs in post-transcriptional gene regulation lies in their ability to modulate gene expression levels with high specificity and efficiency (Roquis et al., 2021). By binding to the 3' untranslated region (UTR) of target mRNAs, miRNAs can trigger mRNA degradation or inhibit translation, preventing the synthesis of specific proteins (Fabian et al., 2010). This fine-tuning of gene expression allows organisms to respond to developmental cues, environmental changes, and stress conditions (Khraiwesh et al., 2012).

n stress responses, small RNAs are crucial components of the adaptive machinery (Shi et al., 2018). When organisms are exposed to various stresses, such as heat, drought, salinity, or pathogen attacks, the expression profiles of small RNAs change dynamically (Li et al., 2010). These stress-responsive small RNAs can target stress-related genes, transcription factors, and signaling components, thereby regulating stress-responsive pathways (Liu and He, 2020). The importance of small RNAs in stress responses is further highlighted by their involvement in cross-talk between different stress signaling pathways (Shukla et al., 2008). Small RNAs can mediate the communication between other stress-responsive genes and pathways, allowing the organism to coordinate its response to multiple stresses simultaneously (Kamthan et al., 2015). Moreover, small RNAs are involved in epigenetic modifications, which can influence gene expression patterns across generations (Zhao et al., 2022). They can guide DNA methylation and histone modifications to specific genomic loci, leading to stable changes in gene expression that can be inherited (Roquis et al., 2021).

Small RNAs are of great importance in post-transcriptional gene regulation and stress responses (Table 1). They provide a versatile and precise mechanism for fine-tuning gene expression, enabling organisms to adapt and respond to changing environments and stressful conditions. Understanding small RNAs' functions and regulatory roles in these processes can provide valuable insights into the molecular mechanisms underlying gene regulation and stress adaptation, with potential applications in various fields, including agriculture, medicine, and biotechnology.

**Table 1** different types of small RNAs and transposable elements involved in regulating stress tolerance in plants

| Regulation of Stress Tolerance in Plants | Type of Small RNA / Transposable Element | Function | Involvement in Stress Tolerance | References |
|---|---|---|---|---|
| **microRNAs (miRNAs)** | Small RNA molecules | Regulate gene expression by targeting messenger RNAs (mRNAs) for degradation or translational repression | Play crucial roles in stress response pathways and modulate the expression of stress-related genes | Zhang (2015) |
| **small interfering RNAs (siRNAs)** | Small RNA molecules | Involved in RNA interference (RNAi) pathway, leading to mRNA degradation or transcriptional gene silencing | Regulate the expression of stress-responsive genes and contribute to stress tolerance | Mallory et al. (2008) |
| **long non-coding RNAs (lncRNAs)** | Longer RNA molecules | Play diverse regulatory roles in gene expression and chromatin organization | Regulate stress-responsive genes, chromatin remodeling, and stress memory in plants | Li et al. (2021) |
| **small nucleolar RNAs (snoRNAs)** | Small RNA molecules | Involved in modification and processing of other RNAs, including ribosomal RNA (rRNA) and small nuclear RNA (snRNA) | Participate in stress response pathways and affect the biogenesis of stress-responsive RNAs | Brown et al. (2003) |
| **trans-acting small interfering RNAs (ta-siRNAs)** | Small RNA molecules | Generated from non-coding TAS precursor transcripts and target mRNAs for cleavage or translational repression | Regulate stress-responsive genes and contribute to stress tolerance | Adenot et al. (2006). |

| | | | | |
|---|---|---|---|---|
| **repeat-associated small interfering RNAs (rasiRNAs)** | Small RNA molecules | Derived from repetitive DNA elements and involved in RNA-directed DNA methylation (RdDM) | Regulate the expression and epigenetic silencing of repetitive elements, which can impact stress tolerance | Wang et al. (2013) |
| **small RNA-derived transposable element small interfering RNAs (te-siRNAs)** | Small RNA molecules derived from TEs | Involved in TE silencing and control the activity and transposition of transposable elements | Play a role in stress adaptation by regulating the expression and mobility of transposable elements | McCue et al. (2015) |
| **long terminal repeat (LTR) retrotransposons** | Transposable elements | DNA elements that can move within the genome through a "copy and paste" mechanism | Can regulate gene expression, contribute to genetic diversity, and impact stress responses in plants | Grandbastien (2015) |
| **CACTA, hAT, and Mutator transposable element families** | Transposable elements | Various types of DNA transposons that can change their position in the genome through a "cut and paste" mechanism | Can induce genetic variation, alter gene expression patterns, and potentially impact stress tolerance in plants | Feschotte and Pritham (2007) |
| **MITE (Miniature Inverted-repeat Transposable Element) and Stowaway transposable elements** | Transposable elements | Short DNA elements that can transpose within the genome and have structural similarities to DNA transposons | Contribute to genetic variation, gene regulation, and stress responses in plants | Bureau and Wessler (1994) |

## Regulation of Small RNAs under Heat Stress

The regulation of small RNAs is critical in the plant's response to elevated temperature (Table 2). Heat stress triggers changes in the expression and activity of various small RNA molecules, such as miRNAs and siRNAs (Ito, 2022). These small RNAs help fine-tune gene expression and orchestrate adaptive responses to cope with heat-induced cellular damage (Wang et al., 2020). During heat stress, specific miRNAs are upregulated or downregulated, resulting in the modulation of target gene expression (Li et al., 2010). These miRNAs can target mRNAs encoding proteins involved in stress response pathways, heat shock proteins, and other factors crucial for plant survival under high temperatures (Li et al., 2010; Liu and He, 2020). By regulating the expression of these target genes, small RNAs contribute to activating stress-responsive pathways and maintaining cellular homeostasis (Matsunaga et al., 2015).

Additionally, heat stress can also impact the biogenesis of small RNAs. The processing of precursor molecules into mature miRNAs and siRNAs may be altered under high temperatures (Yang and Li, 2011). Heat stress can affect the activity of enzymes involved in small RNA biogenesis, leading to changes in the abundance and functionality of these molecules (Zhang et al., 2012). Overall, regulating small RNAs under heat stress is vital for plants to adapt and survive in challenging environmental conditions (Li et al., 2010). Understanding the precise roles and regulatory networks of small RNAs during heat stress can provide insights into the molecular mechanisms underlying plant heat tolerance, which can have implications for crop improvement and agricultural practices in the face of climate change.

**Table 2** different types of reactive oxygen species (ROS) produced in plants under heat stress and the role of different small RNAs and transposable elements in scavenging these ROS to improve tolerance

| ROS Production in plants under heat Stress | Reactive Oxygen Species (ROS) | Small RNAs | Transposable Elements | Role in ROS Scavenging and Heat Stress Tolerance | References |
|---|---|---|---|---|---|
| **Superoxide Radical (O2•-)** | | miR398 | | miR398 regulates the expression of Cu/Zn superoxide dismutase (SOD) genes, enhancing antioxidant defense against superoxide radicals | Sunkar (2010) |
| | | miR398 | | | Fang et al. (2009) |
| | | | Tnt1 Retrotransposon | Tnt1 retrotransposon contributes to enhanced antioxidant defense by increasing the expression of SOD genes | Nicolau et al. (2021) |
| | | | Rider Retrotransposon | Rider retrotransposon regulates the expression of Cu/Zn SOD genes and confers heat stress tolerance | Kaushik et al. (2022) |

| | | | | |
|---|---|---|---|---|
| **Hydrogen Peroxide (H2O2)** | | miR398 | miR398 regulates the expression of Cu/Zn SOD genes, enhancing antioxidant defense against hydrogen peroxide | Sunkar (2010) |
| | miR398 | | | Fang et al. (2009) |
| | miR319 | | miR319 targets TCP transcription factors, enhancing the expression of catalase and peroxidase genes for H2O2 detoxification | Liu et al. (2017) |
| | miR319 | | | Wang et al. (2011) |
| | miR393 | | miR393 targets auxin receptor genes, regulating ROS homeostasis and stress tolerance | Jiao et al. (2010) |
| | miR393 | | | Chen et al. (2011) |
| | miR408 | | miR408 targets plantacyanin-like genes, enhancing peroxidase activity for H2O2 scavenging | Zhang et al. (2017) |
| | miR398 | | | |
| **Singlet Oxygen (1O2)** | miR164 | | miR164 targets NAC transcription factors, enhancing singlet oxygen detoxification | Hewezi et al. (2012) |
| | miR164 | | | Miyazaki et al. (2015) |
| | | Rider Retrotransposon | Rider retrotransposon regulates the expression of peroxidase genes for singlet oxygen detoxification | Kaushik et al. (2022) |

| | | | | |
|---|---|---|---|---|
| **Peroxynitrite (ONOO-)** | miR408 | | miR408 targets plantacyanin-like genes, enhancing peroxidase activity for ONOO- scavenging | Zhang et al. (2017) |
| | miR408 | | | Jin and Wu (2015) |
| | | Tnt1 Retrotransposon | Tnt1 retrotransposon contributes to enhanced antioxidant defense by increasing the expression of peroxidase genes | Nicolau et al. (2021) |
| **Hydroxyl Radical (OH•)** | miR408 | | miR408 targets plantacyanin-like genes, enhancing peroxidase activity for OH• scavenging | Zhang et al. (2017) |
| | miR408 | | | Jin and Wu (2015) |
| | miR394 | | miR394 targets F-box transcription factors, regulating the expression of genes involved in OH• detoxification | Li et al. (2007) |
| | miR394 | | | Debernardi et al., 2012 |
| | | Tnt1 Retrotransposon | Tnt1 retrotransposon contributes to enhanced antioxidant defense by increasing the expression of peroxidase genes | Nicolau et al. (2021) |

## Effects of heat stress on Small RNA Populations and expression patterns

Heat stress profoundly impacts small RNA populations and their expression patterns in various organisms, including plants (Shivaprasad et al., 2012). Small RNAs, such as miRNAs and siRNAs, are essential regulators of gene expression at the post-transcriptional level (Borges and Martienssen, 2015). Under heat stress conditions, the dynamics of small RNA populations undergo significant changes, leading to altered expression patterns and functional consequences (Kruszka et al., 2014). One of the powerful effects of heat stress on small RNA populations is the differential expression of specific miRNAs. Heat stress can lead to upregulating or downregulating specific miRNAs, affecting their target genes' expression (Li et al., 2010). These differentially expressed miRNAs often play critical roles in stress response pathways, cellular homeostasis, and developmental processes (Xin et al., 2010). By fine-tuning the expression of target genes, these miRNAs contribute to the plant's adaptation to heat stress (Mittler et al., 2012).

Heat stress also influences the biogenesis of small RNAs. The processing of precursor molecules into mature miRNAs and siRNAs can be affected under high temperatures (Li et al., 2017). Heat stress alters the activity of enzymes involved in small RNA biogenesis, leading to changes in the abundance and diversity of small RNA populations (Lewis et al., 2015). This can result in generating novel miRNAs or siRNAs and modulating their target gene repertoire (Lewis et al., 2015). Furthermore, heat stress can impact the stability and turnover of small RNAs. High temperatures can accelerate the degradation of small RNAs, reducing their overall abundance and potentially affecting their regulatory functions (Lewis et al., 2015). Additionally, heat stress can alter the association of small RNAs with RNA-binding proteins or other factors involved in their transport and localization within cells (Schiessl et al., 2019).

Overall, the effects of heat stress on small RNA populations and expression patterns have significant implications for gene regulation and stress responses. By modulating the expression of specific miRNAs and influencing small RNA biogenesis, heat stress can reshape the regulatory landscape of the cell, leading to changes in gene expression patterns and cellular responses to heat-induced stress. Understanding these effects is crucial for unravelling the molecular mechanisms underlying heat stress adaptation and developing strategies to enhance heat tolerance in plants and other organisms.

## Influence of salicylic acid and nitric oxide signaling on small RNA biogenesis and activity under heat stress

SA and NO regulate various aspects of plant physiology, including small RNA biogenesis and activity (Ezin and Symonds, 2023; Corpas et al., 2013). Under heat stress conditions, the levels of SA and NO in plants often increase (Tripathi et al., 2021; Rong et al., 2016). These molecules are critical mediators in the heat stress response, coordinating the plant's defense mechanisms (Tripathi et al., 2021; Zhang et al., 2012). SA is known for its involvement in the activation of defense genes and the induction of SAR (Vlot et al., 2009; Park et al., 2013). At the same time, NO is involved in regulating various physiological processes, including plant growth, development, and stress responses (Casson and Hetherington, 2010; Ezin and Symonds, 2023).

Both SA and NO signaling pathways have been found to influence small RNA biogenesis and activity under heat stress. Studies have shown that SA can modulate the expression of critical genes involved in small RNA biogenesis, such as Dicer-like (DCL) and Argonaute (AGO) proteins (Liu and He, 2020; Belide et al., 2012). These proteins process precursor molecules into mature small RNAs and facilitate their target gene silencing. SA can enhance the expression and activity of DCL and AGO proteins, thereby promoting small RNA biogenesis and increasing the abundance of specific small RNA populations (Fang et al., 2015; Abid et al., 2020). Similarly, NO signaling has been shown to impact small RNA biogenesis and activity. NO can regulate gene expression in small RNA biogenesis pathways, including DCL and AGO genes (Zhang et al., 2012; He et al., 2019). Additionally, NO can influence the stability and turnover of small RNAs by modulating their association with RNA-binding proteins or other factors involved in small RNA metabolism (Ezin and Symonds, 2023; He et al., 2019). These regulatory effects of NO contribute to the modulation of small RNA populations and their activity in response to heat stress.

Furthermore, SA and NO signaling pathways can intersect and interact with each other to regulate small RNA biogenesis and activity under heat stress. Studies have demonstrated cross-talk between SA and NO signaling pathways, where SA can stimulate NO production, and NO can enhance SA accumulation (Yang and Huang, 2014; Laubinger, et al., 2010). This cross-talk allows for a coordinated response to heat stress, ensuring optimal small RNA-mediated gene regulation. SA and NO signaling pathways significantly modulate small RNA biogenesis and activity under heat stress. They influence the expression of critical genes involved in small RNA biogenesis and regulate the stability and turnover of small RNAs (Kumar, 2014; Li et al., 2010). The interplay between SA and NO signaling pathways further complicates regulating small RNA-mediated gene expression under heat stress conditions. Understanding these interactions and their impact on small RNA dynamics is crucial for unraveling the molecular mechanisms underlying plant responses to heat stress and developing strategies to enhance crop heat tolerance.

## Cross-talk between small RNA pathways and other heat stress response mechanisms

Small RNA pathways, such as miRNAs and siRNAs, play a crucial role in regulating gene expression and are known to respond to plants' heat stress (Shi et al., 2018). Interestingly, there is a significant cross-talk between small RNA pathways and other heat stress response mechanisms, further enhancing the plant's ability to cope with high temperatures. A critical aspect of the cross-talk between small RNA pathways and heat stress response mechanisms is the regulation of transcription factors (TFs). TFs are crucial regulators of gene expression, and small RNAs can modulate their activity. Under heat stress conditions, specific miRNAs can target and degrade TF transcripts, leading to the downregulation of these TFs. This, in turn, affects the expression of downstream heat stress-responsive genes, allowing the plant to fine-tune its response to high temperatures (Matsunaga et al., 2015).

Small RNAs can also regulate the expression of heat shock proteins (HSPs), which are essential for protecting plants against heat stress (Rai et al., 2020; Sun et al., 2021; Zhang et al., 2020). Heat stress-responsive miRNAs can target and negatively regulate HSP transcripts, reducing HSP expression. This regulation is necessary to balance folding and prevent protein aggregation under heat-stress conditions. Moreover, small RNAs can also target genes involved in protein degradation pathways, such as the ubiquitin-proteasome system, to modulate the turnover of heat stress-related proteins (Zhao et al., 2022). Furthermore, there is evidence of cross-regulation between small RNA and hormonal signaling pathways involved in heat stress responses. Hormones such as abscisic acid (ABA) and auxins are critical in mediating plant responses to heat stress. Small RNAs have been shown to regulate gene expression in ABA and auxin signaling pathways, influencing the plant's hormonal response to heat stress (Verma et al., 2016; Yang and Huang, 2014). Conversely, hormonal signaling pathways can also modulate the expression of small RNAs. For instance, ABA has been reported to enhance the biogenesis of specific miRNAs under heat stress conditions (Matsunaga et al., 2015).

Another level of cross-talk exists between small RNA pathways and epigenetic modifications, such as DNA methylation and histone modifications (Chinnusamy and Zhu, 2009; Liu et al., 2017; Zhang et al., 2023). Heat stress can induce changes in DNA methylation patterns and histone modifications, which can influence the expression of small RNA pathway genes and the activity of small RNAs themselves. In turn, small RNAs can target and guide DNA and histone modifications to regulate gene expression in response to heat stress (Papolu et al., 2022; Ezin and Symonds, 2023). Cross-talk between small RNA pathways and other heat stress response mechanisms is a complex and intricate network that allows plants to finely regulate gene expression in response to high temperatures. This cross-talk involves the regulation of TFs, HSPs, hormonal signaling pathways, and epigenetic modifications (Verma et al., 2016; Li et al., 2010; Laubinger, et al., 2010; Verma et al., 2016). Understanding these interactions will provide valuable insights into the molecular mechanisms underlying plant heat stress responses and may offer new avenues for improving crop heat tolerance.

## Transposable Elements and Heat Stress

TEs are repetitive DNA sequences that can move within a genome and have the potential to disrupt gene function or expression (Doğan and Liu, 2018; Casacuberta and González, 2013). While TEs were traditionally considered "junk DNA," recent studies have revealed their significant impact on genome evolution and gene regulation (Feschotte, 2008; Slotkin and Martienssen, 2007). Interestingly, heat stress has been found to influence TEs and their activity within the genome. Under heat stress conditions, the movement of TEs can be positively and negatively regulated (Papolu et al., 2022; Wigge, 2013). On the one hand, heat stress can lead to the activation of certain TEs, resulting in their increased mobility within the genome. This increased mobility can potentially cause genomic instability and alter the expression of nearby genes (Mirouze and Paszkowski, 2011; Sunkar et al., 2007). On the other hand, heat stress can also induce mechanisms to suppress TE activity. The plant's defense systems may recognize the potentially harmful effects of TEs and activate means to silence or repress their expression (Ito, 2022; Ezin and Symonds, 2023).

One way heat stress influences TEs is through epigenetic modifications. Epigenetic marks, such as DNA methylation and histone modifications, regulate TE activity. Heat stress can induce changes in DNA methylation patterns and histone modifications, which can affect the expression and mobility of TEs (Roquis et al., 2021; Rai et al., 2020). For example, heat stress can lead to DNA demethylation, which may activate certain TEs (Ezin and Symonds, 2023; Laubinger, et al., 2010). Conversely, heat stress can also trigger the deposition of repressive histone marks, such as H3K9me2, to silence TE expression (Belide et al., 2012; Verma et al., 2016).

Furthermore, small RNAs, including miRNAs and siRNAs, also participate in the regulation of TEs under heat stress. Small RNAs can target and guide DNA and histone modifications to specific genomic regions, including TEs. They can induce DNA methylation and histone modifications that suppress TE activity (Zhao et al., 2022; Matsunaga et al., 2015). Conversely, if the small RNA-mediated silencing machinery is compromised under heat stress, the activity of TEs may be derepressed, leading to potential genomic instability (Wang et al., 2020; He et al., 2019). The influence of heat stress on TEs has implications for the genome's stability and adaptability. While the activation of TEs can lead to genomic instability and potentially harmful effects, it can also contribute to genome diversification and adaptation to changing environments. TEs may serve as a source of genetic variation that can facilitate plants' evolution of heat stress tolerance (Ezin and Symonds, 2023; He et al., 2019).

## Activation and regulation of transposable elements under heat stress conditions

The activation and regulation of TEs can undergo significant changes within the genome. TEs are repetitive DNA sequences that can move and insert themselves into different genomic locations, potentially disrupting gene function and causing genomic instability. Heat stress can influence the activity of TEs through various mechanisms. One of the ways heat stresses activates TEs is by inducing DNA damage and genomic instability. High temperatures can lead to DNA double-strand breaks and other forms of DNA damage. These DNA lesions can trigger the activation of TE transposition machinery, resulting in increased TE mobility and transposition events. The disrupted genomic regions can further promote the expression and activity of nearby TEs (Thieme et al., 2017; Zhang et al., 2020).

Another mechanism by which heat stress influences TEs is through alterations in epigenetic regulation. Epigenetic modifications, such as DNA methylation and histone modifications, are crucial in silencing and regulating TEs (Slotkin and Martienssen, 2007). Heat stress can induce changes in DNA methylation patterns, leading to the activation of previously silenced TEs (Mirouze et al., 2009). Additionally, heat stress can affect histone modifications, resulting in changes in chromatin structure and accessibility, which can impact TE expression and mobility (Matsunaga et al., 2015; Liu et al., 2004). Furthermore, heat stress can trigger the production of small RNAs involved in TE regulation. Small RNAs can target and silence TEs at the transcriptional or post-transcriptional level (Ding and Voinnet, 2007; Li et al., 2017). Heat stress can affect the biogenesis and abundance of these small RNAs, leading to altered TE regulation (Borges and Martienssen, 2015; Zhai et al., 2008). For instance, heat stress may reduce the production of siRNAs that target TEs, resulting in their repression (Zhang et al., 2016).

Notably, the regulation of TEs under heat stress conditions is not solely focused on their activation. The genome has evolved defense mechanisms to suppress TE activity and maintain genomic stability. Heat stress can induce the expression of stress-responsive genes, including those involved in TE silencing (Bouyer et al., 2017). These genes encode proteins that recognize and repress TE expression through various mechanisms, such as DNA methylation, histone modifications, and small RNA-mediated pathways (Mirouze et al., 2009; McCue et al., 2015). Heat stress conditions can activate and regulate TEs through multiple mechanisms, including DNA damage-induced activation, changes in epigenetic modifications, small RNA biogenesis, and activation of defense mechanisms (Roquis et al., 2021; Borges and Martienssen, 2015; Liu et al., 2004). The interplay between heat stress and TEs has significant implications for genome stability, evolution, and adaptation to stressful environments (Doğan and Liu, 2018). Further research is needed to unravel the precise molecular mechanisms involved in TE activation and regulation under heat stress and to understand their broader impact on plant biology.

## Significance of transposable elements in heat stress memory and genome stability

TEs play a significant role in plants' heat stress memory and genome stability. Heat stress can induce changes in the genome, and TEs contribute to these alterations and influence the plant's response to subsequent stress events. TEs in heat stress memory are their ability to insert themselves into the genome. TEs can be activated and mobilized during heat stress, leading to their insertion into new genomic locations. These insertions can alter the regulation of nearby genes and impact gene expression patterns (Makarevitch et al., 2015; Quadrana et al., 2016). Some TE insertions can be heritable, meaning they are passed on to subsequent generations. This heritable variation generated by TEs can contribute to the plant's ability to remember and respond more effectively to heat stress in the future (Zhang et al., 2012; Zhao et al., 2022).

TEs also play a role in genome stability during heat stress. Although TEs can cause genomic instability by disrupting gene function, they can also contribute to genome stability through various mechanisms. TEs can act as a source of genetic variation and facilitate the generation of new genetic combinations that may enhance the plant's adaptability to heat stress (Doğan and Liu, 2018; Feschotte, 2008). TEs can also contribute to DNA repair processes by providing sequences that can be used as templates for DNA repair mechanisms (Borges and Martienssen, 2015; Chen et al., 1997).

Furthermore, TEs are involved in regulating stress-responsive genes and signaling pathways. They can influence the expression of stress-related genes by inserting themselves into their regulatory regions or by generating small RNAs that target these genes (Makarevitch et al., 2015; Quadrana et al., 2016). This regulation mediated by TEs can impact the plant's ability to withstand heat stress and maintain cellular homeostasis.

Importantly, TEs are not solely passive participants in heat stress responses; the plant can actively regulate them. The plant's defense mechanisms include the repression of TEs to maintain genome stability. Heat stress can trigger the activation of specific genes and pathways involved in TE silencing and regulation. These mechanisms help to suppress the potentially harmful effects of TEs on the genome (Dai et al., 2018; Marino-Ramírez et al., 2005). Transposable elements have a dual role in heat stress responses. They can contribute to genome instability by causing genetic disruptions but also play a crucial role in heat stress memory and genome stability (Pavelka et al., 2010; Quadrana et al., 2016). The heritable changes induced by TEs can enhance the plant's adaptive capacity, allowing it to respond more effectively to future stress events. Understanding the significance of TEs in heat stress responses and genome stability can provide valuable insights into plant adaptation strategies and aid in developing crops with improved stress tolerance.

## Molecular Mechanisms of Small RNA-Mediated Regulation of Transposable Elements

Small RNAs regulate TEs through various molecular mechanisms. One mechanism involves siRNAs derived from TE transcripts. These siRNAs guide a protein complex known as RISC to target and degrade TE transcripts, preventing their expression (Abid et al., 2020; Kutter et al., 2007). Another mechanism involves miRNAs targeting TE transcripts by base-pairing with complementary sequences. This binding inhibits TE transcript translation or promotes their degradation (Li et al., 2017; Cai et al., 2019). Additionally, miRNAs can regulate the expression of host genes that control TE activity, indirectly influencing TE regulation (Cui et al., 2017; Nadarajah and Kumar, 2019). Furthermore, small RNAs can trigger DNA methylation, an epigenetic modification that can silence TEs. The siRNAs or miRNAs can guide DNA methylation machinery to TE sequences, adding methyl groups to the DNA and effectively silencing the TEs (Slotkin and Martienssen, 2021; Cuerda-Gil and Slotkin, 2016).

Small RNAs also contribute to TE regulation through "transitive silencing." In this process, siRNAs produced from an active TE can spread to other TEs with similar sequences, leading to their silencing (Brown et al., 2012). This spreading of silencing signals helps to maintain genome stability and control TE activity. Small RNA-mediated regulation of TEs involves siRNAs and miRNAs that can target TE transcripts for degradation or translational repression, induce DNA methylation to silence TEs, and facilitate transitive silencing to control the activity of related TEs (Kanno et al., 2010; Li et al., 2017; Zhai et al., 2008). These molecular mechanisms form a complex regulatory network that ensures the proper control and suppression of TEs in the genome, contributing to genome stability and adequate functioning of the host organism.

## Insights into the molecular mechanisms underlying small RNA-mediated regulation of transposable elements

Small RNA-mediated regulation of TEs involves a complex molecular mechanism network that ensures genome stability and proper gene expression. Small RNAs, including siRNAs and miRNAs play crucial roles in this regulation. SiRNAs are generated from double-stranded RNA (dsRNA) precursors, which can arise from TE transcripts or the formation of RNA hairpin structures within TE sequences (Roquis et al., 2021; Borges and Martienssen, 2015). These siRNAs guide the RISC to complementary sequences in TE transcripts. Once bound, RISC cleaves the TE transcripts, leading to their degradation and preventing their translation (Llave et al., 2002; Plasterk, 2002). This process, known as post-transcriptional gene silencing, effectively represses TE activity.

MiRNAs, on the other hand, are small non-coding RNAs that are transcribed from their genes. They can also target TE transcripts by base-pairing with complementary sequences. This interaction can result in translational repression or degradation of the TE transcripts (Axtell, 2013; Zhang et al., 2020). Also, miRNAs can indirectly regulate TEs by targeting host genes controlling TE activity. By modulating the expression of these host genes, miRNAs can influence the overall activity of TEs in the genome (Zhang et al., 2016). Another important mechanism of small RNA-mediated TE regulation is DNA methylation. SiRNAs and miRNAs can guide DNA methyltransferases to TE sequences, adding methyl groups to the DNA. This methylation results in gene silencing by preventing the binding of transcription factors and other regulatory proteins to the TE sequences (Debernardi et al., 2012; Law and Jacobsen, 2010). DNA methylation-mediated silencing is a heritable form of TE regulation and helps maintain genome stability.

Furthermore, small RNAs can participate in a phenomenon called transitive silencing. In this process, siRNAs generated from an active TE can spread to other TEs with similar sequences, leading to their silencing. This amplification of silencing signals helps control the activity of multiple TEs and maintains genome integrity. Molecular mechanisms underlying small RNA-mediated regulation of TEs involve siRNAs and miRNAs that target TE transcripts for degradation or translational repression, guide DNA methylation to silence TEs, and facilitate transitive silencing to control the activity of related TEs. These mechanisms provide a sophisticated regulatory network that adequately suppresses TEs and preserves genome stability in organisms.

## Role of RNA interference (RNAi) machinery and RNA-induced silencing complex (RISC) in transposable element control

RNA interference (RNAi) machinery and the RISC play crucial roles in the control of transposable elements (TEs), which are mobile genetic elements that can move within a genome (Baulcombe, 2004). The RNAi pathway is a conserved mechanism in various organisms, including plants and animals, that regulates gene expression at the post-transcriptional level. The process begins with generating small RNA molecules, such as siRNAs, critical effectors of RNAi. siRNAs are derived from longer dsRNA molecules, which can be formed from TE transcripts or the formation of RNA hairpin structures within TE sequences. The dsRNA is processed by an enzyme called Dicer, which cleaves it into short siRNA duplexes (Kim et al., 2009).

The siRNA duplexes are then loaded onto the RISC, which consists of the Argonaute protein and other associated proteins. The siRNA duplex is unwound within the RISC, generating a single-stranded siRNA molecule. One strand of the siRNA, the guide strand, is retained in the RISC, while the other, called the passenger strand, is degraded (Hannon, 2002). The guide siRNA strand in the RISC then binds to complementary sequences within TE transcripts. This interaction leads to the recruitment of additional proteins, such as endonucleases, which cleave the TE transcripts at specific sites. This cleavage event triggers the degradation of the TE RNA, preventing its translation into functional transposase enzymes responsible for TE mobility. As a result, the TE is effectively silenced, reducing its activity and potential impact on genome stability (Qi et al., 2006).

The RISC mediates the cleavage and degradation of TE transcripts and also participates in the establishment of epigenetic modifications, such as DNA methylation, which can lead to long-term repression of TEs. The guide siRNAs can direct DNA methyltransferases to TE sequences, adding methyl groups to the DNA. This DNA methylation contributes to the stable silencing of TEs by preventing the binding of transcription factors and other regulatory proteins, thereby blocking their transcription and reducing mobility (Matzke et al., 2015; Law and Jacobsen, 2010)

In summary, the RNAi machinery and RISC play essential roles in the control of transposable elements. TE transcripts are selectively targeted for degradation or epigenetic modification by generating siRNAs and loading them onto the RISC, effectively repressing their activity and maintaining genome stability. This intricate regulatory system helps organisms maintain proper control over transposable elements, potential sources of genomic instability and genetic variation.

## Importance of small RNA-guided DNA methylation and chromatin modifications in transposable element regulation

Small RNA-guided DNA methylation and chromatin modifications are crucial in regulating TEs, which are mobile genetic elements in genomes. TEs can potentially cause genomic instability and disrupt gene expression if left uncontrolled. To counteract this threat, organisms have evolved mechanisms to silence and regulate TEs, and small RNA-guided DNA methylation and chromatin modifications are key players in these processes. Small RNAs, such as siRNAs, are generated from TE transcripts or derived from the formation of RNA hairpin structures within TE sequences. These siRNAs guide the recruitment of an RISC to complementary TE sequences. Once bound, the RISC triggers a series of events that lead to the establishment of DNA methylation and chromatin modifications (Slotkin and Martienssen, 2007; Brown et al., 2012).

One of the significant effects of small RNA-guided regulation is the recruitment of DNA methyltransferases (DNMTs) to TE sequences. DNMTs catalyze the addition of methyl groups to the DNA, resulting in DNA methylation. This modification can lead to the stable repression of TEs by inhibiting their transcription and preventing the binding of transcription factors and other regulatory proteins. DNA methylation provides a heritable epigenetic mark that can be maintained through DNA replication, ensuring long-term silencing of TEs across cell generations (Law and Jacobsen, 2010). In addition to DNA methylation, small RNA-guided regulation also influences chromatin modifications. Chromatin is the complex of DNA and proteins that make up chromosomes, and its structure can affect gene expression. The binding of small RNAs to TE sequences can recruit proteins that modify chromatin, such as histone methyltransferases and histone deacetylases. These enzymes add or remove chemical groups from histone proteins associated with DNA in the chromatin structure. The resulting changes in histone modifications can lead to the compaction or relaxation of chromatin, making it more or less accessible for transcription factors and other regulatory proteins. This, in turn, affects the expression of nearby genes, including TEs (Feng et al., 2010).

Small RNA-guided DNA methylation and chromatin modifications in TE regulation provide stable and heritable control over these mobile genetic elements. By silencing TEs and preventing their activation, organisms can maintain genome stability and protect against potentially harmful effects of TE mobility, such as insertional mutagenesis or disruption of essential genes. These mechanisms ensure the integrity and proper functioning of the genome, allowing organisms to adapt and survive in dynamic environments.

## Conclusion

The regulation of small RNAs and transposable elements is crucial for plant responses to heat stress. SA and NO are essential in this regulatory process. SA and NO act as signaling molecules that modulate gene expression and contribute to plant stress tolerance. Under heat stress conditions, SA and NO influence each other's signaling pathways, leading to a synergistic effect on regulating small RNAs and transposable elements. The accumulation of small RNAs suppresses transposable element activity, ensuring genome stability and preventing potential damage caused by transposon mobilization. The interplay between SA and NO pathways involves complex molecular interactions and transcriptional regulation mechanisms that fine-tune the expression of small RNAs and transposable elements. Understanding these regulatory networks provides valuable insights for developing strategies to enhance plant resilience to heat stress and improve crop productivity in challenging environments. By manipulating the SA and NO signaling pathways and harnessing the power of small RNA-mediated regulation, it is possible to enhance plant stress tolerance and mitigate the detrimental effects of heat stress on crop production. Further research in this area will contribute to developing innovative approaches for sustainable agriculture in the face of increasing global temperatures and climate change challenges.

## References

Abid, G., Ouertani, R.N., Jebara, S.H., Boubakri, H., Muhovski, Y., Ghouili, E., Abdelkarim, S., Chaieb, O., Hidri, Y., Kadri, S. and El Ayed, M., 2020. Alleviation of drought stress in faba bean (*Vicia faba* L.) by exogenous application of β-aminobutyric acid (BABA). Physiology and Molecular Biology of Plants, 26, pp.1173-1186.

Adenot, X., Elmayan, T., Lauressergues, D., Boutet, S., Bouché, N., Gasciolli, V. and Vaucheret, H., 2006. DRB4-dependent TAS3 trans-acting siRNAs control leaf morphology through AGO7. Current Biology, 16(9), pp.927-932.

Axtell, M.J. and Bartel, D.P., 2005. Antiquity of microRNAs and their targets in land plants. The Plant Cell, 17(6), pp.1658-1673.

Axtell, M.J. and Bowman, J.L., 2008. Evolution of plant microRNAs and their targets. Trends in plant science, 13(7), pp.343-349.

Axtell, M.J., 2013. Classification and comparison of small RNAs from plants. Annual review of plant biology, 64, pp.137-159.

Bartel, D.P., 2009. MicroRNAs: target recognition and regulatory functions. cell, 136(2), pp.215-233.

Baulcombe, D., 2004. RNA silencing in plants. Nature, 431(7006), pp.356-363.

Belide, S., Petrie, J.R., Shrestha, P. and Singh, S.P., 2012. Modification of seed oil composition in Arabidopsis by artificial microRNA-mediated gene silencing. Frontiers in plant science, 3, p.168.

Borges, F. and Martienssen, R.A., 2015. The expanding world of small RNAs in plants. Nature reviews Molecular cell biology, 16(12), pp.727-741.

Bouyer, D., Kramdi, A., Kassam, M., Heese, M., Schnittger, A., Roudier, F. and Colot, V., 2017. DNA methylation dynamics during early plant life. Genome biology, 18(1), pp.1-12.

Brown, J.D., Mitchell, S.E. and O'neill, R.J., 2012. Making a long story short: noncoding RNAs and chromosome change. Heredity, 108(1), pp.42-49.

Brown, J.W., Echeverria, M. and Qu, L.H., 2003. Plant snoRNAs: functional evolution and new modes of gene expression. Trends in plant science, 8(1), pp.42-49.

Bureau, T.E. and Wessler, S.R., 1994. Mobile inverted-repeat elements of the Tourist family are associated with the genes of many cereal grasses. Proceedings of the National Academy of Sciences, 91(4), pp.1411-1415.

Cai, H., Yang, C., Liu, S., Qi, H., Wu, L., Xu, L.A. and Xu, M., 2019. MiRNA-target pairs regulate adventitious rooting in Populus: a functional role for miR167a and its target Auxin response factor 8. Tree physiology, 39(11), pp.1922-1936.

Carthew, R.W. and Sontheimer, E.J., 2009. Origins and mechanisms of miRNAs and siRNAs. Cell, 136(4), pp.642-655.

Casacuberta, E. and González, J., 2013. The impact of transposable elements in environmental adaptation. Molecular ecology, 22(6), pp.1503-1517.

Casson, S.A. and Hetherington, A.M., 2010. Environmental regulation of stomatal development. Current opinion in plant biology, 13(1), pp.90-95.

Chen, M., SanMiguel, P., De Oliveira, A.C., Woo, S.S., Zhang, H., Wing, R.A. and Bennetzen, J.L., 1997. Microcolinearity in sh2-homologous regions of the maize, rice, and sorghum genomes. Proceedings of the National Academy of Sciences, 94(7), pp.3431-3435.

Chen, Z.H., Bao, M.L., Sun, Y.Z., Yang, Y.J., Xu, X.H., Wang, J.H., Han, N., Bian, H.W. and Zhu, M.Y., 2011. Regulation of auxin response by miR393-targeted transport inhibitor response protein 1 is involved in normal development in Arabidopsis. Plant molecular biology, 77, pp.619-629.

Chinnusamy, V. and Zhu, J.K., 2009. Epigenetic regulation of stress responses in plants. Current opinion in plant biology, 12(2), pp.133-139.

Corpas, F.J., Alché, J.D. and Barroso, J.B., 2013. Current overview of S-nitrosoglutathione (GSNO) in higher plants. Frontiers in plant science, 4, p.126.

Cuerda-Gil, D. and Slotkin, R.K., 2016. Non-canonical RNA-directed DNA methylation. Nature plants, 2(11), pp.1-8.

Cui, X. and Cao, X., 2014. Epigenetic regulation and functional exaptation of transposable elements in higher plants. Current Opinion in Plant Biology, 21, pp.83-88.

Dai, X., Zhuang, Z. and Zhao, P.X., 2018. psRNATarget: a plant small RNA target analysis server (2017 release). Nucleic acids research, 46(W1), pp.W49-W54.

Debernardi, J.M., Rodriguez, R.E., Mecchia, M.A. and Palatnik, J.F., 2012. Functional specialization of the plant miR396 regulatory network through distinct microRNA–target interactions. PLoS genetics, 8(1), p.e1002419.

Ding, S.W. and Voinnet, O., 2007. Antiviral immunity directed by small RNAs. Cell, 130(3), pp.413-426.

Doğan, E.S. and Liu, C., 2018. Three-dimensional chromatin packing and positioning of plant genomes. Nature Plants, 4(8), pp.521-529.

Ezin, V. and Symonds, R.C., 2023. MicroRNA-mediated Regulation of Heat Stress Response. Plant MicroRNAs and Stress Response, pp.90-119.

Fabian, M.R., Sonenberg, N. and Filipowicz, W., 2010. Regulation of mRNA translation and stability by microRNAs. Annual review of biochemistry, 79, pp.351-379.

Fang, Y., Liao, K., Du, H., Xu, Y., Song, H., Li, X. and Xiong, L., 2015. A stress-responsive NAC transcription factor SNAC3 confers heat and drought tolerance through modulation of reactive oxygen species in rice. Journal of experimental botany, 66(21), pp.6803-6817.

Fang, Z., Shao, C., Meng, Y., Wu, P. and Chen, M., 2009. Phosphate signaling in Arabidopsis and *Oryza sativa*. Plant Science, 176(2), pp.170-180.

Feng, S., Cokus, S.J., Zhang, X., Chen, P.Y., Bostick, M., Goll, M.G., Hetzel, J., Jain, J., Strauss, S.H., Halpern, M.E. and Ukomadu, C., 2010. Conservation and divergence of methylation patterning in plants and animals. Proceedings of the National Academy of Sciences, 107(19), pp.8689-8694.

Feschotte, C. and Pritham, E.J., 2007. DNA transposons and the evolution of eukaryotic genomes. Annu. Rev. Genet., 41, pp.331-368.

Feschotte, C., 2008. Transposable elements and the evolution of regulatory networks. Nature Reviews Genetics, 9(5), pp.397-405.

Grandbastien, M.A., 2015. LTR retrotransposons, handy hitchhikers of plant regulation and stress response. Biochimica et Biophysica Acta (BBA)-Gene Regulatory Mechanisms, 1849(4), pp.403-416.

Ha, M. and Kim, V.N., 2014. Regulation of microRNA biogenesis. Nature reviews Molecular cell biology, 15(8), pp.509-524.

Hannon, G.J., 2002. RNA interference. nature, 418(6894), pp.244-251.

He, J., Jiang, Z., Gao, L., You, C., Ma, X., Wang, X., Xu, X., Mo, B., Chen, X. and Liu, L., 2019. Genome-wide transcript and small RNA profiling reveals transcriptomic responses to heat stress. Plant physiology, 181(2), pp.609-629.

Hewezi, T., Maier, T.R., Nettleton, D. and Baum, T.J., 2012. The Arabidopsis microRNA396-GRF1/GRF3 regulatory module acts as a developmental regulator in the reprogramming of root cells during cyst nematode infection. Plant physiology, 159(1), pp.321-335.

Huang, L.Z., Zhou, M., Ding, Y.F. and Zhu, C., 2022. Gene networks involved in plant heat stress response and tolerance. International Journal of Molecular Sciences, 23(19), p.11970.

Ito, H., 2022. Environmental stress and transposons in plants. Genes & Genetic Systems, 97(4), pp.169-175.

Jiao, Y., Wang, Y., Xue, D., Wang, J., Yan, M., Liu, G., Dong, G., Zeng, D., Lu, Z., Zhu, X. and Qian, Q., 2010. Regulation of OsSPL14 by OsmiR156 defines ideal plant architecture in rice. Nature genetics, 42(6), pp.541-544.

Jin, W. and Wu, F., 2015. Identification and characterization of cucumber microRNAs in response to Pseudoperonospora cubensis infection. Gene, 569(2), pp.225-232.

Kalendar, R., et al. (2010). Mobile inverted-repeat elements of the Tourist family are associated with the genes of many cereal grasses.

Kamthan, A., Chaudhuri, A., Kamthan, M. and Datta, A., 2015. Small RNAs in plants: recent development and application for crop improvement. Frontiers in plant science, 6, p.208.

Kanno, T., Bucher, E., Daxinger, L., Huettel, B., Kreil, D.P., Breinig, F., Lind, M., Schmitt, M.J., Simon, S.A., Gurazada, S.G.R. and Meyers, B.C., 2010. RNA-directed DNA methylation and plant development require an IWR1-type transcription factor. EMBO reports, 11(1), pp.65-71.

Kaushik, V., Tiwari, M. and Tiwari, V., 2022. Interaction of RecA mediated SOS response with bacterial persistence, biofilm formation, and host response. International Journal of Biological Macromolecules.

Khraiwesh, B., Zhu, J.K. and Zhu, J., 2012. Role of miRNAs and siRNAs in biotic and abiotic stress responses of plants. Biochimica et Biophysica Acta (BBA)-Gene Regulatory Mechanisms, 1819(2), pp.137-148.

Kim, V.N., Han, J. and Siomi, M.C., 2009. Biogenesis of small RNAs in animals. Nature reviews Molecular cell biology, 10(2), pp.126-139.

Kruszka, K., Pacak, A., Swida-Barteczka, A., Nuc, P., Alaba, S., Wroblewska, Z., Karlowski, W., Jarmolowski, A. and Szweykowska-Kulinska, Z., 2014. Transcriptionally and post-transcriptionally regulated microRNAs in heat stress response in barley. Journal of experimental botany, 65(20), pp.6123-6135.

Kumar, R., 2014. Role of microRNAs in biotic and abiotic stress responses in crop plants. Applied biochemistry and biotechnology, 174, pp.93-115.

Kutter, C., Schöb, H., Stadler, M., Meins Jr, F. and Si-Ammour, A., 2007. MicroRNA-mediated regulation of stomatal development in Arabidopsis. The Plant Cell, 19(8), pp.2417-2429.

Laubinger, S., Zeller, G., Henz, S.R., Buechel, S., Sachsenberg, T., Wang, J.W., Rätsch, G. and Weigel, D., 2010. Global effects of the small RNA biogenesis machinery on the Arabidopsis thaliana transcriptome. Proceedings of the National Academy of Sciences, 107(41), pp.17466-17473.

Law, J.A. and Jacobsen, S.E., 2010. Establishing, maintaining and modifying DNA methylation patterns in plants and animals. Nature Reviews Genetics, 11(3), pp.204-220.

Lewis, L.A., Polanski, K., de Torres-Zabala, M., Jayaraman, S., Bowden, L., Moore, J., Penfold, C.A., Jenkins, D.J., Hill, C., Baxter, L. and Kulasekaran, S., 2015. Transcriptional dynamics driving MAMP-triggered immunity and pathogen effector-mediated immunosuppression in Arabidopsis leaves following infection with Pseudomonas syringae pv tomato DC3000. The Plant Cell, 27(11), pp.3038-3064.

Li, H., Sun, J., Xu, Y., Jiang, H., Wu, X. and Li, C., 2007. The bHLH-type transcription factor AtAIB positively regulates ABA response in Arabidopsis. Plant molecular biology, 65, pp.655-665.

Li, S., Castillo-González, C., Yu, B. and Zhang, X., 2017. The functions of plant small RNA s in development and in stress responses. The Plant Journal, 90(4), pp.654-670.

Li, Y.F., Zheng, Y., Addo-Quaye, C., Zhang, L., Saini, A., Jagadeeswaran, G., Axtell, M.J., Zhang, W. and Sunkar, R., 2010. Transcriptome-wide identification of microRNA targets in rice. The Plant Journal, 62(5), pp.742-759.

Li, Z., Tian, P., Huang, T. and Huang, J., 2021. Noncoding-RNA-mediated regulation in response to macronutrient stress in plants. International Journal of Molecular Sciences, 22(20), p.11205.

Liu, J. and He, Z., 2020. Small DNA methylation, big player in plant abiotic stress responses and memory. Frontiers in Plant Science, 11, p.595603.

Liu, J., Cheng, X., Liu, P., Li, D., Chen, T., Gu, X. and Sun, J., 2017. MicroRNA319-regulated TCPs interact with FBHs and PFT1 to activate CO transcription and control flowering time in Arabidopsis. PLoS genetics, 13(5), p.e1006833.

Liu, J., Feng, J., & Han, X. (2022). Heat stress-responsive retrotransposon activation and its potential impact on plant genome evolution and phenotypic variation. Plant Communications, 3(2), 100254. doi: 10.1016/j.xplc.2022.100254.

Liu, J., He, Y., Amasino, R. and Chen, X., 2004. siRNAs targeting an intronic transposon in the regulation of natural flowering behavior in Arabidopsis. Genes & development, 18(23), pp.2873-2878.

Liu, L., Sonbol, F.M., Huot, B., Gu, Y., Withers, J., Mwimba, M., Yao, J., He, S.Y. and Dong, X., 2016. Salicylic acid receptors activate jasmonic acid signaling through a non-canonical pathway to promote effector-triggered immunity. Nature communications, 7(1), p.13099.

Liu, Q., Yan, S., Yang, T., Zhang, S., Chen, Y.Q. and Liu, B., 2017. Small RNAs in regulating temperature stress response in plants. Journal of Integrative Plant Biology, 59(11), pp.774-791.

Liu, S., de Jonge, J., Trejo-Arellano, M.S., Santos-González, J., Köhler, C. and Hennig, L., 2021. Role of H1 and DNA methylation in selective regulation of transposable elements during heat stress. New Phytologist, 229(4), pp.2238-2250.

Llave, C., Kasschau, K.D., Rector, M.A. and Carrington, J.C., 2002. Endogenous and silencing-associated small RNAs in plants. The Plant Cell, 14(7), pp.1605-1619.

Makarevitch, I., Waters, A.J., West, P.T., Stitzer, M., Hirsch, C.N., Ross-Ibarra, J. and Springer, N.M., 2015. Correction: transposable elements contribute to activation of maize genes in response to abiotic stress. PLoS genetics, 11(10), p.e1005566.

Mallory, A.C., Reinhart, B.J., Jones-Rhoades, M.W., Tang, G., Zamore, P.D., Barton, M.K. and Bartel, D.P., 2004. MicroRNA control of PHABULOSA in leaf development: importance of pairing to the microRNA 5′ region. The EMBO journal, 23(16), pp.3356-3364.

Mao, H., Wang, H., Liu, S., Li, Z., Yang, X., Yan, J., Li, J., Tran, L.S.P. and Qin, F., 2015. A transposable element in a NAC gene is associated with drought tolerance in maize seedlings. Nature communications, 6(1), p.8326.

Mariño-Ramírez, L., Kann, M.G., Shoemaker, B.A. and Landsman, D., 2005. Histone structure and nucleosome stability. Expert review of proteomics, 2(5), pp.719-729.

Matsunaga, W., Ohama, N., Tanabe, N., Masuta, Y., Masuda, S., Mitani, N., Yamaguchi-Shinozaki, K., Ma, J.F., Kato, A. and Ito, H., 2015. A small RNA mediated regulation of a stress-activated retrotransposon and the tissue specific transposition during the reproductive period in Arabidopsis. Frontiers in Plant Science, 6, p.48.

Matzke, M.A., Kanno, T. and Matzke, A.J., 2015. RNA-directed DNA methylation: the evolution of a complex epigenetic pathway in flowering plants. Annual review of plant biology, 66, pp.243-267.

McCue, A.D., Panda, K., Nuthikattu, S., Choudury, S.G., Thomas, E.N. and Slotkin, R.K., 2015. ARGONAUTE 6 bridges transposable element m RNA-derived si RNA s to the establishment of DNA methylation. The EMBO journal, 34(1), pp.20-35.

Meng, Y., Jing, H., Huang, J., Shen, R. and Zhu, X., 2022. The role of nitric oxide signaling in plant responses to cadmium stress. International Journal of Molecular Sciences, 23(13), p.6901.

Mirouze, M. and Paszkowski, J., 2011. Epigenetic contribution to stress adaptation in plants. Current opinion in plant biology, 14(3), pp.267-274.

Mirouze, M., Reinders, J., Bucher, E., Nishimura, T., Schneeberger, K., Ossowski, S., Cao, J., Weigel, D., Paszkowski, J. and Mathieu, O., 2009. Selective epigenetic control of retrotransposition in Arabidopsis. Nature, 461(7262), pp.427-430.

Mittler, R., Finka, A. and Goloubinoff, P., 2012. How do plants feel the heat?. Trends in biochemical sciences, 37(3), pp.118-125.

Miura, K. and Tada, Y., 2014. Regulation of water, salinity, and cold stress responses by salicylic acid. Frontiers in plant science, 5, p.4.

Miyazaki, Y., Abe, H., Takase, T., Kobayashi, M. and Kiyosue, T., 2015. Overexpression of LOV KELCH PROTEIN 2 confers dehydration tolerance and is associated with enhanced expression of dehydration-inducible genes in *Arabidopsis thaliana*. Plant cell reports, 34, pp.843-852.

Nadarajah, K. and Kumar, I.S., 2019. Drought response in rice: The miRNA story. International journal of molecular sciences, 20(15), p.3766.

Nicolau, M., Picault, N. and Moissiard, G., 2021. The evolutionary volte-face of transposable elements: From harmful jumping genes to major drivers of genetic innovation. Cells, 10(11), p.2952.

Papolu, P.K., Ramakrishnan, M., Mullasseri, S., Kalendar, R., Wei, Q., Zou, L., Ahmad, Z., Vinod, K.K., Yang, P. and Zhou, M., 2022. Retrotransposons: How the continuous evolutionary front shapes plant genomes for response to heat stress. Frontiers in plant science, 13, p.1064847.

Park, S.W., Li, W., Viehhauser, A., He, B., Kim, S., Nilsson, A.K., Andersson, M.X., Kittle, J.D., Ambavaram, M.M., Luan, S. and Esker, A.R., 2013. Cyclophilin 20-3 relays a 12-oxo-phytodienoic acid signal during stress responsive regulation of cellular redox homeostasis. Proceedings of the National Academy of Sciences, 110(23), pp.9559-9564.

Pavelka, N., Rancati, G., Zhu, J., Bradford, W.D., Saraf, A., Florens, L., Sanderson, B.W., Hattem, G.L. and Li, R., 2010. Aneuploidy confers quantitative proteome changes and phenotypic variation in budding yeast. Nature, 468(7321), pp.321-325.

Plasterk, R.H., 2002. RNA silencing: the genome's immune system. Science, 296(5571), pp.1263-1265.

Prakash, V., Singh, V.P., Tripathi, D.K., Sharma, S. and Corpas, F.J., 2021. Nitric oxide (NO) and salicylic acid (SA): A framework for their relationship in plant development under abiotic stress. Plant Biology, 23, pp.39-49.

Qi, Y., He, X., Wang, X.J., Kohany, O., Jurka, J. and Hannon, G.J., 2006. Distinct catalytic and non-catalytic roles of ARGONAUTE4 in RNA-directed DNA methylation. Nature, 443(7114), pp.1008-1012.

Quadrana, L., Bortolini Silveira, A., Mayhew, G.F., LeBlanc, C., Martienssen, R.A., Jeddeloh, J.A. and Colot, V., 2016. The Arabidopsis thaliana mobilome and its impact at the species level. elife, 5, p.e15716.

Rai, K.K., Pandey, N. and Rai, S.P., 2020. Salicylic acid and nitric oxide signaling in plant heat stress. Physiologia plantarum, 168(2), pp.241-255.

Rong, D., Luo, N., Mollet, J.C., Liu, X. and Yang, Z., 2016. Salicylic acid regulates pollen tip growth through an NPR3/NPR4-independent pathway. Molecular plant, 9(11), pp.1478-1491.

Roquis, D., Robertson, M., Yu, L., Thieme, M., Julkowska, M. and Bucher, E., 2021. Genomic impact of stress-induced transposable element mobility in Arabidopsis. Nucleic acids research, 49(18), pp.10431-10447.

Schiessl, S.V., Quezada-Martinez, D., Tebartz, E., Snowdon, R.J. and Qian, L., 2019. The vernalisation regulator FLOWERING LOCUS C is differentially expressed in biennial and annual *Brassica napus*. Scientific reports, 9(1), p.14911.

Shi, G.Q., Fu, J.Y., Rong, L.J., Zhang, P.Y., Guo, C.J. and Kai, X.I.A.O., 2018. TaMIR1119, a miRNA family member of wheat (*Triticum aestivum*), is essential in the regulation of plant drought tolerance. Journal of integrative agriculture, 17(11), pp.2369-2378.

Shivaprasad, P.V., Chen, H.M., Patel, K., Bond, D.M., Santos, B.A. and Baulcombe, D.C., 2012. A microRNA superfamily regulates nucleotide binding site–leucine-rich repeats and other mRNAs. The Plant Cell, 24(3), pp.859-874.

Shivaraj, S.M., Vats, S., Bhat, J.A., Dhakte, P., Goyal, V., Khatri, P., Kumawat, S., Singh, A., Prasad, M., Sonah, H. and Sharma, T.R., 2020. Nitric oxide and hydrogen sulfide crosstalk during heavy metal stress in plants. Physiologia Plantarum, 168(2), pp.437-455.

Shukla, L.I., Chinnusamy, V. and Sunkar, R., 2008. The role of microRNAs and other endogenous small RNAs in plant stress responses. Biochimica et Biophysica Acta (BBA)-Gene Regulatory Mechanisms, 1779(11), pp.743-748.

Siddiqui, M.H., Al-Whaibi, M.H. and Basalah, M.O., 2011. Role of nitric oxide in tolerance of plants to abiotic stress. Protoplasma, 248, pp.447-455.

Slotkin, R.K. and Martienssen, R., 2007. Transposable elements and the epigenetic regulation of the genome. Nature reviews genetics, 8(4), pp.272-285.

Sun, X., Huang, N., Li, X., Zhu, J., Bian, X., Li, H., Wang, L., Hu, Q. and Luo, H., 2021. A chloroplast heat shock protein modulates growth and abiotic stress response in creeping bentgrass. Plant, Cell & Environment, 44(6), pp.1769-1787.

Sunkar, R., 2010, October. MicroRNAs with macro-effects on plant stress responses. In Seminars in cell & developmental biology (Vol. 21, No. 8, pp. 805-811). Academic Press.

Sunkar, R., Chinnusamy, V., Zhu, J. and Zhu, J.K., 2007. Small RNAs as big players in plant abiotic stress responses and nutrient deprivation. Trends in plant science, 12(7), pp.301-309.

Thieme, M., Lanciano, S., Balzergue, S., Daccord, N., Mirouze, M. and Bucher, E., 2017. Inhibition of RNA polymerase II allows controlled mobilisation of retrotransposons for plant breeding. Genome Biology, 18, pp.1-10.

Tripathi, D.K., Vishwakarma, K., Singh, V.P., Prakash, V., Sharma, S., Muneer, S., Nikolic, M., Deshmukh, R., Vaculik, M. and Corpas, F.J., 2021. Silicon crosstalk with reactive oxygen species, phytohormones and other signaling molecules. Journal of hazardous materials, 408, p.124820.

Verma, V., Ravindran, P. and Kumar, P.P., 2016. Plant hormone-mediated regulation of stress responses. BMC plant biology, 16, pp.1-10.

Vlot, A.C., Dempsey, D.M.A. and Klessig, D.F., 2009. Salicylic acid, a multifaceted hormone to combat disease. Annual review of phytopathology, 47, pp.177-206.

Wang, H.Q., Sun, L.P., Wang, L.X., Fang, X.W., Li, Z.Q., Zhang, F.F., Hu, X., Qi, C. and He, J.M., 2020. Ethylene mediates salicylic-acid-induced stomatal closure by controlling reactive oxygen species and nitric oxide production in Arabidopsis. Plant Science, 294, p.110464.

Wang, J.W., Park, M.Y., Wang, L.J., Koo, Y., Chen, X.Y., Weigel, D. and Poethig, R.S., 2011. miRNA control of vegetative phase change in trees. PLoS genetics, 7(2), p.e1002012.

Wang, X., Duan, C.G., Tang, K., Wang, B., Zhang, H., Lei, M., Lu, K., Mangrauthia, S.K., Wang, P., Zhu, G. and Zhao, Y., 2013. RNA-binding protein regulates plant DNA methylation by controlling mRNA processing at the intronic heterochromatin-containing gene IBM1. Proceedings of the National Academy of Sciences, 110(38), pp.15467-15472.

Wigge, P.A., 2013. Ambient temperature signaling in plants. Current opinion in plant biology, 16(5), pp.661-666.

Xin, M., Wang, Y., Yao, Y., Xie, C., Peng, H., Ni, Z. and Sun, Q., 2010. Diverse set of microRNAs are responsive to powdery mildew infection and heat stress in wheat (Triticum aestivum L.). BMC plant biology, 10(1), pp.1-11.

Yadav, C.B., Muthamilarasan, M., Pandey, G. and Prasad, M., 2015. Identification, characterization and expression profiling of Dicer-like, Argonaute and RNA-dependent RNA polymerase gene families in foxtail millet. Plant molecular biology reporter, 33, pp.43-55.

Yang, L. and Huang, H., 2014. Roles of small RNAs in plant disease resistance. Journal of integrative plant biology, 56(10), pp.962-970.

Yang, X. and Li, L., 2011. miRDeep-P: a computational tool for analyzing the microRNA transcriptome in plants. Bioinformatics, 27(18), pp.2614-2615.

Zhai, J., Liu, J., Liu, B., Li, P., Meyers, B.C., Chen, X. and Cao, X., 2008. Small RNA-directed epigenetic natural variation in Arabidopsis thaliana. PLoS genetics, 4(4), p.e1000056.

Zhai, J., Zhang, H., Arikit, S., Huang, K., Nan, G.L., Walbot, V. and Meyers, B.C., 2015. Spatiotemporally dynamic, cell-type–dependent premeiotic and meiotic phasiRNAs in maize anthers. Proceedi

Zhang, B. and Wang, Q., 2015. MicroRNA-based biotechnology for plant improvement. Journal of cellular physiology, 230(1), pp.1-15.

Zhang, J.P., Yu, Y., Feng, Y.Z., Zhou, Y.F., Zhang, F., Yang, Y.W., Lei, M.Q., Zhang, Y.C. and Chen, Y.Q., 2017. MiR408 regulates grain yield and photosynthesis via a phytocyanin protein. Plant Physiology, 175(3), pp.1175-1185.

Zhang, J.P., Yu, Y., Feng, Y.Z., Zhou, Y.F., Zhang, F., Yang, Y.W., Lei, M.Q., Zhang, Y.C. and Chen, Y.Q., 2017. MiR408 regulates grain yield and photosynthesis via a phytocyanin protein. Plant Physiology, 175(3), pp.1175-1185.

Zhang, X., Xia, J., Lii, Y.E., Barrera-Figueroa, B.E., Zhou, X., Gao, S., Lu, L., Niu, D., Chen, Z., Leung, C. and Wong, T., 2012. Genome-wide analysis of plant nat-siRNAs reveals insights into their distribution, biogenesis and function. Genome biology, 13, pp.1-16.

Zhang, X., Xu, W., Ni, D., Wang, M. and Guo, G., 2020. Genome-wide characterization of tea plant (*Camellia sinensis*) Hsf transcription factor family and role of CsHsfA2 in heat tolerance. BMC Plant Biology, 20(1), pp.1-17.

Zhang, Z., Liu, X., Guo, X., Wang, X.J. and Zhang, X., 2016. Arabidopsis AGO3 predominantly recruits 24-nt small RNAs to regulate epigenetic silencing. Nature Plants, 2(5), pp.1-7.

Zhao, Y., Wei, Q., Chen, T., Xu, L., Liu, J., Zhang, X., Han, G. and Ma, Q., 2022. Identification and characterization of heat-responsive miRNAs and their regulatory network in maize. Plant Growth Regulation, pp.1-14.

# CHAPTER 9

# SALICYLIC ACID AND NITRIC OXIDE IN PLANT HEAT STRESS RESPONSE: STATUS AND WAY AHEAD


## Abstract

Salicylic acid (SA) and nitric oxide (NO) are crucial signaling molecules involved in the response of plants to various stresses, including heat, drought, and pathogen attacks. The cross-talk between SA and NO signaling pathways has emerged as an essential regulatory mechanism in plant stress responses. This cross-talk involves complex molecular interactions and reciprocal regulation between SA and NO, leading to synergistic effects on gene expression and stress tolerance. SA is a critical regulator of defence responses and can modulate the expression of heat stress-responsive genes, small RNAs, and transposable elements, thereby contributing to plant resilience under heat-stress conditions. On the other hand, NO serves as a signaling molecule that induces heat stress tolerance and regulates the expression of heat stress-responsive genes and physiological processes such as stomatal closure and ROS scavenging. The interplay between SA and NO signaling pathways enables plants to mount an effective defence response and adapt to adverse environmental conditions. However, there are still unresolved questions and gaps in our understanding of the specific molecular mechanisms underlying the SA-NO cross-talk.

Further research is needed to unravel the precise signaling pathways, transcriptional regulation networks, and downstream target genes influenced by SA and NO. Technological advancements and emerging approaches, such as omics technologies, provide promising avenues for studying the roles of SA and NO in stress responses and developing strategies to enhance plant stress tolerance. Harnessing the potential of SA and NO in crop improvement offers exciting opportunities for enhancing stress resilience and ensuring sustainable agriculture in the face of changing climatic conditions.

**Keywords**: Salicylic acid, Nitric oxide, Cross-talk, Stress responses, Plant resilience

## Introduction

Salicylic acid (SA) and nitric oxide (NO) are two essential signaling molecules that play crucial roles in the response of plants to heat stress (Bandurska, 2013; Rai et al., 2021; Hara et al., 2012). Heat stress is a significant environmental factor that can negatively impact plant growth, development, and productivity (Wahid et al., 2007; Mittler et al., 2012). Understanding the involvement of SA and NO in heat stress response is essential for developing strategies to enhance plant resilience and stress tolerance (Rai et al., 2021; Hara et al., 2012). Salicylic acid is a well-known signaling molecule involved in defense responses against various stresses, including heat stress (Rai et al., 2021; Hara et al., 2012). It accumulates in plants upon exposure to heat stress conditions and is a crucial regulator of heat stress-responsive genes (Kang et al., 2014; Rivas-San Vicente et al., 2011). SA is known to regulate the expression of genes involved in heat shock protein production, antioxidant defense systems, and other stress-related pathways (Gupta et al., 2011; Hara et al., 2012). It helps plants withstand heat stress by activating defense mechanisms and reducing oxidative damage (Rai et al., 2021; Rivas-San Vicente et al., 2011). Additionally, SA has been found to influence the expression of small RNAs (sRNAs) involved in gene regulation and transposable element control under heat stress conditions (Zhu et al., 2020; Singh et al., 2022).

Nitric oxide is another crucial signaling molecule that participates in heat stress response in plants (Gupta et al., 2011; Chaki et al., 2021; Singhal et al., 2021). Under heat stress conditions, plants produce increased levels of NO, which serves as a signaling molecule to induce heat stress tolerance (Li et al., 2016; Blokhina and Fagerstedt, 2010). NO regulates heat stress-responsive genes encoding heat shock proteins, antioxidants, and signaling components (Wang et al., 2015; Prakash et al., 2020). It helps plants adapt to high temperatures by modulating various physiological processes, including stomatal closure, photosynthesis, and scavenging reactive oxygen species (ROS) (Zheng et al., 2015; Gupta et al., 2011). NO also influences the expression and activity of sRNAs, which contribute to the regulation of transposable elements (TEs) and maintain genome stability under heat stress conditions (Kang et al., 2014; Ohama et al., 2017).

The interplay between SA and NO signaling pathways in the context of heat stress response is an emerging area of research (Rai et al., 2021; Arora et al., 2016). Evidence of cross-talk between these two signaling molecules suggests a synergistic effect on gene expression regulation (Christou et al., 2014; Rai et al., 2020). SA and NO can influence each other's signaling pathways and interact with various transcription factors and other regulatory components to modulate the heat stress response (Wang et al., 2015; Kumar, 2014). Moving forward, further research is needed to unravel the specific molecular mechanisms underlying the actions of SA and NO in plant heat stress response. This includes understanding the precise signaling pathways, transcriptional regulation networks, and downstream target genes influenced by SA and NO (Zhang et al., 2011; Horváth et al., 2007). Additionally, investigating the interactions between SA, NO, sRNAs, and other signaling molecules will provide deeper insights into the intricate regulatory networks involved in heat stress response (Borrelli et al., 2021; Proietti et al., 2013). Overall, the importance of SA and NO in plant heat stress response is evident. Expanding our knowledge of their roles and mechanisms of action will pave the way for developing effective strategies to enhance plant tolerance to heat stress, ultimately contributing to sustainable agriculture and food security in a changing climate.

## Salicylic Acid and Nitric Oxide Cross-talk

SA and NO are two essential signaling molecules in plants that exhibit cross-talk and interact in various physiological processes (Table 1) (Prakash et al., 2020; Zheng et al., 2010). Both SA and NO play crucial roles in plant defense responses against biotic and abiotic stresses, including pathogen attacks, drought, and heat stress (Zheng et al., 2015; Borrelli et al., 2021; Zhou et al., 2021). The cross-talk between SA and NO involves complex molecular interactions and regulatory mechanisms (Wang et al., 2015; Horváth et al., 2007). SA can positively influence NO production, while NO can enhance SA accumulation (Prakash et al., 2020; Wendehenne et al., 2004). This reciprocal regulation between SA and NO enables plants to mount an effective defense response by activating defense genes, inducing antioxidant systems, and regulating various signaling pathways (Kumar et al., 2013; Singhal et al., 2021; Borrelli et al., 2021). The SA-NO cross-talk is an essential component of plant stress signaling and provides an integrated mechanism for plants to coordinate their responses to different stressors (Zhu et al., 2020; Kumar, 2014). Understanding the molecular mechanisms underlying the SA-NO cross-talk will contribute to developing strategies to enhance plant stress tolerance and improve crop productivity in challenging environments (Rai et al., 2020; Proietti et al., 2013).

**Table 1** cross-talk of salicylic acid (SA) and nitric oxide (NO) on plant growth, development, and stress responses

| Cross-talk of Salicylic Acid (SA) and Nitric Oxide (NO) | Aspect | Salicylic Acid (SA) | Nitric Oxide (NO) | Cross-talk Mechanisms and Effects | References |
|---|---|---|---|---|---|
| **Seed Germination** | Germination Rate | SA inhibits seed germination by interacting with ABA signaling pathways and reducing the sensitivity of seeds to germination-promoting factors. | NO promotes seed germination by enhancing embryo cell elongation and endosperm weakening. | SA and NO act antagonistically in seed germination, with SA delaying germination while NO promotes it. | Wang et al. (2015) |
| | Seed Dormancy | SA promotes seed dormancy by interacting with ABA and gibberellic acid (GA) signaling pathways, inhibiting germination. | NO breaks seed dormancy by regulating the expression of genes involved in ABA and GA metabolism and signaling. | SA and NO have opposing effects on seed dormancy, with SA promoting dormancy while NO breaks it. | Prakash et al. (2020) |
| **Plant Growth and Development** | Root Development | SA inhibits primary root elongation and lateral root formation by modulating auxin transport and signaling. | NO promotes primary root growth and lateral root development by enhancing auxin signaling and inducing cell elongation. | SA and NO have antagonistic effects on root development, with SA inhibiting while NO promoting root growth. | Beligni and Lamattina, (2001) |

| | | | | | |
|---|---|---|---|---|---|
| | Shoot Development | SA regulates shoot development by affecting leaf expansion, shoot branching, and flowering time. | NO influences shoot development by promoting leaf growth, shoot branching, and flowering. | SA and NO have synergistic or antagonistic effects on shoot development depending on the specific processes and stages of development. | Prakash et al. (2020) |
| | Flowering Time | SA delays flowering by interacting with photoperiod and gibberellin signaling pathways. | NO promotes flowering by regulating the expression of flowering-related genes and interacting with gibberellin and photoperiod pathways. | SA and NO have contrasting effects on flowering time regulation, with SA delaying flowering while NO promotes it. | Zheng et al., (2010) |
| | Senescence | SA accelerates leaf senescence by promoting the expression of senescence-associated genes. | NO delays leaf senescence by inhibiting the breakdown of chlorophyll and maintaining chloroplast integrity. | SA and NO have opposing effects on leaf senescence, with SA promoting senescence while NO retards it. | Zhang et al. (2011) |
| **Plant Stress Responses** | Abiotic Stress Tolerance | SA enhances abiotic stress tolerance by regulating the expression of stress-related genes, activating antioxidant defense systems, and inducing stomatal closure. | NO improves abiotic stress tolerance by regulating the expression of stress-responsive genes, enhancing antioxidant capacity, and regulating stomatal behavior. | SA and NO synergistically enhance abiotic stress tolerance through the activation of common downstream signaling pathways. | Rai et al. (2021) |

| | | | | | |
|---|---|---|---|---|---|
| | Biotic Stress Responses | SA plays a crucial role in plant defense against pathogens by inducing systemic acquired resistance (SAR) and activating defense-related genes. | NO functions as a signaling molecule in plant defense responses by regulating defense gene expression and promoting hypersensitive response (HR). | SA and NO act cooperatively in plant defense responses, with SA priming the defense system and NO amplifying the defense response. | Khan et al. (2015) |
| | Oxidative Stress | SA acts as an antioxidant by scavenging reactive oxygen species (ROS) and inducing the expression of antioxidant enzymes. | NO functions as a signaling molecule in oxidative stress responses and regulates the expression of antioxidant enzymes. | SA and NO synergistically mitigate oxidative stress by activating antioxidant defense mechanisms and regulating ROS homeostasis. | Li et al. (2019) |
| | Heat Stress | SA plays a crucial role in heat stress tolerance by regulating heat shock protein expression, maintaining membrane stability, and activating antioxidant systems. | NO enhances heat stress tolerance by regulating heat shock protein expression, stabilizing membranes, and modulating antioxidant defense. | SA and NO cooperate in improving heat stress tolerance by activating common heat stress-responsive genes and antioxidant defense mechanisms. | Wang et al. (2015) |

**Table 2** interplay of salicylic acid (SA) and nitric oxide (NO) with other signaling molecules and pathways for comprehensive stress adaptation in plants

| Interplay of SA and NO with Other Signaling Molecules and Pathways | Signaling Molecule/Pathway | Salicylic Acid (SA) | Nitric Oxide (NO) | Interplay Mechanisms and Effects | References |
|---|---|---|---|---|---|
| **Abscisic Acid (ABA)** | ABA Signaling Pathway | SA enhances ABA signaling by upregulating the expression of ABA biosynthesis and signaling genes. | NO interacts with ABA signaling by regulating ABA biosynthesis and signaling components. | SA and NO synergistically enhance ABA signaling, leading to improved stress tolerance and stomatal regulation. | Gupta et al. (2011) |
| | ABA-Responsive Genes | SA induces the expression of ABA-responsive genes involved in stress tolerance, such as RD22, RD26, and DREB2A. | NO regulates the expression of ABA-responsive genes, including RD29A, RD29B, and COR47. | SA and NO cooperate in activating ABA-responsive genes, leading to enhanced stress adaptation. | Zhang et al. (2011) |

| | | | | | |
|---|---|---|---|---|---|
| | Stomatal Regulation | SA promotes stomatal closure by modulating ion channels and regulating the activity of guard cells. | NO mediates stomatal opening by activating the plasma membrane H+-ATPase and regulating ion fluxes. | SA and NO have opposing effects on stomatal regulation, with SA promoting closure and NO promoting opening. | Neill et al. (2002) |
| | Stomatal Sensitivity | SA reduces stomatal sensitivity to ABA, resulting in improved stomatal regulation under stress conditions. | NO enhances stomatal sensitivity to ABA, leading to improved stomatal responses to stress. | SA and NO interact in modulating stomatal sensitivity to ABA, contributing to fine-tuned stomatal responses. | Wang et al. (2015) |
| | ABA Crosstalk with Other Hormones | SA interacts with ABA in crosstalk with other hormones, such as jasmonic acid (JA) and ethylene (ET), to regulate stress responses. | NO participates in ABA crosstalk with other hormones, including JA and ET, to fine-tune stress adaptation. | SA, NO, and ABA collaborate in hormone crosstalk to orchestrate complex stress responses in plants. | Li et al. (2019) |

| | | | | | |
|---|---|---|---|---|---|
| **Reactive Oxygen Species (ROS)** | ROS Signaling Pathway | SA regulates ROS production and scavenging systems, including antioxidant enzymes and non-enzymatic antioxidants. | NO participates in ROS signaling by modulating ROS production and the activities of antioxidant enzymes. | SA and NO coordinate ROS signaling to maintain redox homeostasis, protect against oxidative damage, and regulate stress responses. | Rai et al. (2021) |
| | ROS-Responsive Genes | SA induces the expression of ROS-responsive genes, such as APX, CAT, and GST, involved in antioxidant defense. | NO regulates the expression of ROS-responsive genes, including APX, CAT, and SOD, to enhance antioxidant capacity. | SA and NO synergistically activate ROS-responsive genes, leading to improved antioxidant defense against stress-induced ROS. | Fancy et al. (2017) |
| | ROS-Induced Cell Death | SA mitigates ROS-induced cell death by regulating cell death-related genes and activating anti-apoptotic mechanisms. | NO plays a dual role in ROS-induced cell death, acting as a cell death promoter or protector depending on the cellular context. | SA and NO interact in modulating ROS-induced cell death, fine-tuning cell death pathways under stress conditions. | Wang et al. (2015) |

| | | | | | |
|---|---|---|---|---|---|
| | Antioxidant Enzymes | SA enhances the activities of antioxidant enzymes, including SOD, CAT, and APX, to scavenge ROS and alleviate oxidative stress. | NO regulates the activities of antioxidant enzymes, such as SOD, CAT, and APX, to maintain redox homeostasis and alleviate oxidative stress. | SA and NO collaborate in activating antioxidant enzymes, reinforcing antioxidant defense mechanisms against ROS-mediated damage. | Arasimowicz-Jelonek et al. (2014) |
| **Calcium Signaling** | $Ca^{2+}$ Signaling Pathway | SA modulates $Ca^{2+}$ signaling by regulating the activities of $Ca^{2+}$ channels and $Ca^{2+}$ sensors. | NO modulates $Ca^{2+}$ signaling by regulating the activities of $Ca^{2+}$ channels and $Ca^{2+}$ sensors. | SA and NO collaborate in modulating $Ca^{2+}$ signaling, resulting in enhanced stress responses and adaptive growth. | Li et al. (2019) |
| **MAPK Signaling** | MAPK Signaling Pathway | SA activates MAPK signaling cascades, including MPK3 and MPK6, to regulate stress-responsive gene expression. | NO activates MAPK signaling cascades, such as MPK3 and MPK6, to mediate stress responses. | SA and NO synergistically activate MAPK signaling, leading to coordinated regulation of stress-responsive gene expression. | Gupta et al. (2011) |

| | | | | | |
|---|---|---|---|---|---|
| **Phytohormones** | Crosstalk with Other Hormones | SA crosstalks with other phytohormones, including jasmonic acid (JA) and ethylene (ET), to regulate stress responses. | NO crosstalks with other phytohormones, such as JA and ET, to fine-tune stress adaptation. | SA and NO collaborate in hormone crosstalk, integrating multiple signaling pathways for comprehensive stress adaptation. | Gupta et al. (2011) |
| **Kinase Cascades** | Crosstalk with Kinases | SA interacts with various kinases, such as protein kinases, MAPKs, and receptor-like kinases, to regulate stress responses. | NO interacts with kinases, including protein kinases and MAPKs, to modulate stress signaling pathways. | SA and NO cross-regulate kinase cascades, leading to the activation of stress-responsive signaling pathways. | Fancy et al. (2017) |
| **Epigenetic Regulation** | Epigenetic Modifications | SA regulates DNA methylation and histone modifications to modulate stress-responsive gene expression. | NO participates in DNA methylation and histone modifications to influence gene expression under stress conditions. | SA and NO collaboratively regulate epigenetic modifications, contributing to the fine-tuning of stress-responsive gene expression. | Li et al. (2019) |

| | | | | | |
|---|---|---|---|---|---|
| **Transcription Factors** | Transcriptional Regulation | SA regulates the activity of various transcription factors, including WRKY, NAC, and MYB, to modulate stress-responsive gene expression. | NO modulates the activity of transcription factors, such as WRKY, NAC, and MYB, to regulate stress-responsive gene expression. | SA and NO synergistically regulate the activity of transcription factors, enabling precise control of stress-responsive gene expression. | Khan et al. (2015) |

## Interactions and cross-talk between salicylic acid and nitric oxide signaling pathways

SA is a well-known signaling molecule in plant defense against various stresses, including pathogens and abiotic factors (Vlot et al., 2009). It activates defense genes, promotes the synthesis of antimicrobial compounds, and enhances the production of ROS to combat stress (Beckers and Spoel, 2005; Hara et al., 2012). SA also acts as a critical regulator in systemic acquired resistance (SAR), a long-lasting defense response that protects plants against various pathogens (Vlot et al., 2009). On the other hand, NO is a gaseous signaling molecule that participates in multiple physiological processes, including plant defense and stress responses (Table 2) (Mur et al., 2013). NO is involved in plant growth regulation, stomatal movement, and the modulation of cellular processes through protein modifications (Neill et al., 2002). Under stress conditions, NO acts as a signaling mediator, triggering adaptive responses to enhance stress tolerance (Kumar et al., 2013).

The interactions and cross-talk between SA and NO signaling pathways are complex and context-dependent. In some cases, SA can stimulate NO production, while in other instances, NO can induce the accumulation of SA (Fancy et al., 2017; Mata and Lamattina, 2001). This reciprocal regulation suggests a synergistic relationship between the two molecules, leading to enhanced defense responses. Studies have shown that SA and NO can positively regulate each other's signaling pathways. SA can activate NO synthesis by upregulating the expression of nitric oxide synthase (NOS) genes or stimulating the activity of NOS enzymes (Mata and Lamattina, 2001; Verma et al., 2020). In turn, NO can enhance SA accumulation by promoting the expression of SA biosynthesis-related genes (Borrelli et al., 2021). This mutual reinforcement between SA and NO amplifies plants' defense responses and stress tolerance.

Additionally, SA and NO signaling pathways converge on standard downstream components, such as mitogen-activated protein kinases (MAPKs) and transcription factors, mediating defense-related gene expression (Prakash et al., 2020; Hou and Tsuda, 2022). The activation of these shared components facilitates the integration of SA and NO signals, leading to coordinated defense responses against stressors. The interactions and cross-talk between SA and NO signaling pathways are crucial in plant defense and stress responses (Wang et al., 2015; Singhal et al., 2021). Understanding the molecular mechanisms underlying their interactions will provide valuable insights into regulating plant immunity and stress tolerance. This knowledge can be harnessed to develop strategies for crop improvement and sustainable agriculture in the face of changing environmental conditions (Rai et al., 2021).

## Signaling molecules and components involved in the cross-regulation between salicylic acid and nitric oxide

The cross-regulation between SA and NO signaling pathways involves a complex network of signaling molecules and components. These molecules and elements act as mediators in the cross-talk between SA and NO, influencing their signaling pathways and modulating plant responses to various stresses (Arora et al., 2016; Delaney et al., 1994). One of the key players in the cross-regulation between SA and NO is the enzyme NOS. NOS catalyzes the production of NO from arginine and SA (Chen et al., 2015). SA can regulate its activity and has been shown to enhance NOS expression and activity, increasing NO production in plants (Prakash et al., 2020; Rai et al., 2018). Conversely, NO can also stimulate the expression of SA biosynthesis-related genes, resulting in elevated SA levels (Chen et al., 2015; Zhang et al., 2006). This reciprocal regulation between SA and NO production forms a positive feedback loop, amplifying their signaling cascades.

Another critical component involved in cross-regulation is the MAPK pathway. Both salicylic acid (SA) and nitric oxide (NO) signaling pathways can activate MAPKs, key cellular signaling regulators. MAPKs act as downstream components of SA and NO signaling, transmitting the signals to target molecules such as transcription factors. Activation of MAPKs by SA or NO leads to the phosphorylation and activation of downstream transcription factors, which regulate the expression of defense-related genes (Dubiella et al., 2013; Oz et al., 2015). This convergence of SA and NO signaling pathways on MAPKs provides a mechanism for cross-regulation and integration of their signaling outputs. In addition, ROS play a significant role in the cross-regulation between SA and NO. Both SA and NO can induce the production of ROS as part of their signaling responses. ROS act as secondary messengers, participating in signal transduction and amplification. SA-induced ROS production can activate NO synthesis, while NO-derived ROS can enhance SA accumulation (Wang et al., 2015; Pardo-Hernández et al., 2020). The interplay between ROS, SA, and NO further contributes to the cross-regulation and coordination of defense responses.

Furthermore, several other signaling molecules, such as cyclic guanosine monophosphate (cGMP) and calcium ions ($Ca^{2+}$), have been implicated in the cross-talk between SA and NO (Neill et al., 2002; Verma et al., 2020). SA and NO can influence the levels of cGMP and $Ca^{2+}$, which modulate downstream signaling events and gene expression (Joudoi et al., 2013; Arasimowicz and Floryszak-Wieczorek, 2007). The cross-regulation between SA and NO signaling pathways involves a network of signaling molecules and components. These include NOS, MAPKs, ROS, cGMP, and $Ca^{2+}$ (Zhang et al., 2006; Wang et al., 2015; Verma et al., 2020). Understanding the interplay and interactions among these components is essential for unraveling the complex regulatory mechanisms underlying the cross-regulation between SA and NO, ultimately leading to a comprehensive understanding of plant defense and stress responses.

## Impact of Cross-talk on plant growth, development, and stress responses

The cross-talk between different signaling pathways, including the interaction between SA and NO, significantly impacts plant growth, development, and stress responses (Rai et al., 2021; Borrelli et al., 2021). The intricate communication between these pathways allows plants to fine-tune their physiological and biochemical processes to adapt to changing environmental conditions (Li et al., 2019). One of the significant impacts of the cross-talk between SA and NO is on plant defense responses. SA is well-known for activating defense mechanisms against pathogens and pests, while NO acts as a signaling molecule in both biotic and abiotic stress responses (Choudhury et al., 2017; Singhal et al., 2021). The interaction between SA and NO signaling pathways leads to a synergistic effect on plant defense (Delaney et al., 1994). The combined action of SA and NO enhances the production of defense-related compounds, such as pathogenesis-related proteins and secondary metabolites, and strengthens the plant's ability to combat stressors (Zheng et al., 2015; Zhang et al., 2011). This cross-talk also promotes the activation of antioxidant systems, which helps mitigate oxidative stress caused by various environmental factors (Ahmad et al., 2016).

Furthermore, the cross-talk between SA and NO influences plant growth and development processes. SA has been shown to regulate various aspects of plant growth, including seed germination, root development, and flowering (Rai et al., 2021; Delangiz et al., 2022; Horváth et al., 2007). On the other hand, NO plays a crucial role in regulating plant growth and development, such as seedling growth, root elongation, and leaf expansion (Rai et al., 2020; Borrelli et al., 2021; Kolbert et al., 2019). The interplay between SA and NO signaling pathways modulates these growth processes, ensuring a balanced and coordinated development of plants (Delaney et al., 1994; Arora et al., 2016). The cross-talk between SA and NO pathways also impacts plant responses to abiotic stresses, including heat, drought, and salt stress. Both SA and NO have been implicated in enhancing stress tolerance and promoting plant survival under adverse conditions (Du et al., 2021a; Rai et al., 2018; Hayat et al., 2007). Their interaction leads to the activation of stress-responsive genes and the modulation of stress-related physiological processes (Khan et al., 2015; Proietti et al., 2013). The combined action of SA and NO helps plants to cope with stress-induced damage, maintain cellular homeostasis, and improve their overall stress tolerance (Kumar, 2014; Arora et al., 2016).

## Salicylic Acid and Nitric Oxide in Stress Responses

SA and NO are signaling molecules essential in plant stress responses. SA is primarily associated with defense against biotic stresses, such as pathogens and pests. It triggers a cascade of defense responses, including activating defense genes, producing antimicrobial compounds, and strengthening the plant cell wall (Vlot et al., 2009; Fu and Dong, 2013; Pieterse et al., 2012). SA also regulates SAR, a long-lasting defense mechanism that protects against many pathogens (Durrant and Dong, 2004; Yan and Dong, 2014). On the other hand, NO is involved in both biotic and abiotic stress responses. It acts as a signaling molecule that regulates various physiological processes, including plant growth, development, and stress tolerance (Hancock, 2020; Blokhina and Fagerstedt, 2010). NO helps plants cope with abiotic stresses like drought, high salinity, and temperature extremes. It promotes antioxidant activity, regulates stomatal closure, and enhances the expression of stress-responsive genes (Gupta et al., 2011; Arasimowicz-Jelonek & Floryszak-Wieczorek, 2014; Khan et al., 2015).

Interestingly, SA and NO often interact and influence each other's signaling pathways, forming a cross-talk that fine-tunes plant stress responses. Depending on the specific stress conditions and cellular context, they can act synergistically or antagonistically. The cross-talk between SA and NO involves complex molecular interactions, including the modulation of enzyme activities, gene expression, and protein modifications (Mur et al.,

2013; Verma et al., 2020). Overall, the interaction between SA and NO is critical for integrating different stress signaling pathways and coordinating plant responses to both biotic and abiotic stresses. Understanding the intricate interplay between SA and NO signaling pathways will provide insights into how plants perceive and respond to stress and may contribute to developing strategies to enhance stress tolerance in crops (Zheng et al., 2015; Gupta et al., 2011).

## Role of salicylic acid and nitric oxide in abiotic stress tolerance, including drought, heat, and salt stress

SA and NO are essential signaling molecules that play crucial roles in plant responses to abiotic stresses, including drought, heat, and salt stress. These stresses pose significant challenges to plant growth and development, and understanding the mechanisms by which SA and NO contribute to abiotic stress tolerance is of great importance for improving crop productivity. Drought stress is characterized by a lack of water availability, reducing plant growth and productivity. SA and NO have been shown to enhance drought tolerance in plants. SA regulates stomatal closure, reducing water loss through transpiration and maintaining cellular water balance. It also promotes the accumulation of osmoprotectants, such as proline and soluble sugars, which help plants cope with water deficit (Janda et al., 2020; Rai et al., 2020; Borrelli et al., 2021). Similarly, NO plays a role in regulating stomatal closure and reducing water loss. It also enhances the expression of stress-related genes involved in osmotic adjustment and antioxidant defense, thereby improving plant survival under drought conditions (Wang et al., 2015; Rai et al., 2020; Oz et al., 2015).

Heat stress is another significant abiotic stress that adversely affects plant growth and development. SA and NO have been demonstrated to confer heat stress tolerance in plants. SA accumulation under heat stress conditions triggers the expression of heat shock proteins (HSPs) that help protect cellular structures and maintain protein homeostasis (Kang et al., 2014; Rivas-San Vicente et al., 2011). SA also regulates antioxidant defense mechanisms, reducing oxidative damage caused by heat stress (Nazar et al., 2015). On the other hand, NO acts as a signaling molecule that induces the expression of heat stress-responsive genes, including HSPs, and enhances plant thermotolerance (Rai et al., 2020; Pardo-Hernández et al., 2020).

Salt stress, resulting from high soil salinity, leads to ion imbalances and osmotic stress in plants. SA and NO play essential roles in mediating salt stress tolerance. SA helps maintain ion homeostasis by regulating ion transporters and improving the efficiency of ion exclusion from sensitive tissues (Rai et al., 2021; Hayat et al., 2007). It also promotes the accumulation of compatible solutes, which help maintain osmotic balance (Herrera-Vásquez et al., 2015). NO enhances salt stress tolerance by improving ion uptake and compartmentalization and regulating antioxidant defenses against salt-induced oxidative stress (Rai et al., 2018; Arora et al., 2016).

## Interplay with other signaling molecules and pathways for comprehensive stress adaptation

The interplay between different signaling molecules and pathways is crucial for comprehensive stress adaptation in plant stress responses. SA, NO, and other signaling molecules (Figure 1) coordinate plant responses and enhance stress tolerance. One crucial interaction occurs between SA and jasmonic acid (JA), two key signaling molecules involved in defense responses. SA is typically associated with defense against biotrophic pathogens, while JA protects against necrotrophic pathogens and herbivores. However, there can be a cross-talk between the SA and JA signaling pathways under certain stress conditions, such as drought or high salinity. This cross-talk allows plants to fine-tune their defense responses and prioritize the most appropriate defense strategy based on the specific stress encountered (Meena et al., 2017, 2020; Pieterse et al., 2012; Van der Does et al., 2013).

Additionally, the interaction between SA and abscisic acid (ABA) is critical for plant responses to abiotic stresses, particularly drought and high salinity. ABA is a hormone that regulates stomatal closure and helps plants conserve water during water deficit conditions. SA can enhance ABA signaling, increasing stomatal closure and improving water use efficiency. This interaction between SA and ABA ensures that plants can effectively manage water resources and survive stressful conditions (Jiang et al., 1996; Wang et al., 2015).

Furthermore, the interplay between SA and ROS is essential for plant stress adaptation. ROS, such as hydrogen peroxide ($H_2O_2$), are generated as byproducts of various cellular processes, including stress responses. SA can modulate ROS levels and regulate their signaling pathways. It acts as a positive regulator of antioxidant enzymes that scavenge ROS and prevent oxidative damage. This interplay between SA and ROS helps plants maintain redox homeostasis and minimize cellular damage caused by stress-induced ROS accumulation (Dubiella et al., 2013; Miura & Tada, 2014).

Moreover, the cross-talk between NO and other signaling molecules, such as ABA and ROS, is critical for plant stress responses. NO can modulate ABA signaling and influence stomatal closure, thus regulating water loss and improving drought tolerance. It also interacts with ROS to regulate oxidative stress and antioxidant defense pathways. NO acts as a signaling molecule that activates antioxidant enzymes and enhances their activity, thereby mitigating oxidative damage caused by stress. These interactions allow for fine-tuning defense responses, regulation of stomatal closure and water use efficiency, modulation of ROS levels, and activation of antioxidant defense mechanisms (Arasimowicz-Jelonek & Floryszak-Wieczorek, 2014; Beligni and Lamattina, 2001; Khan et al., 2015).

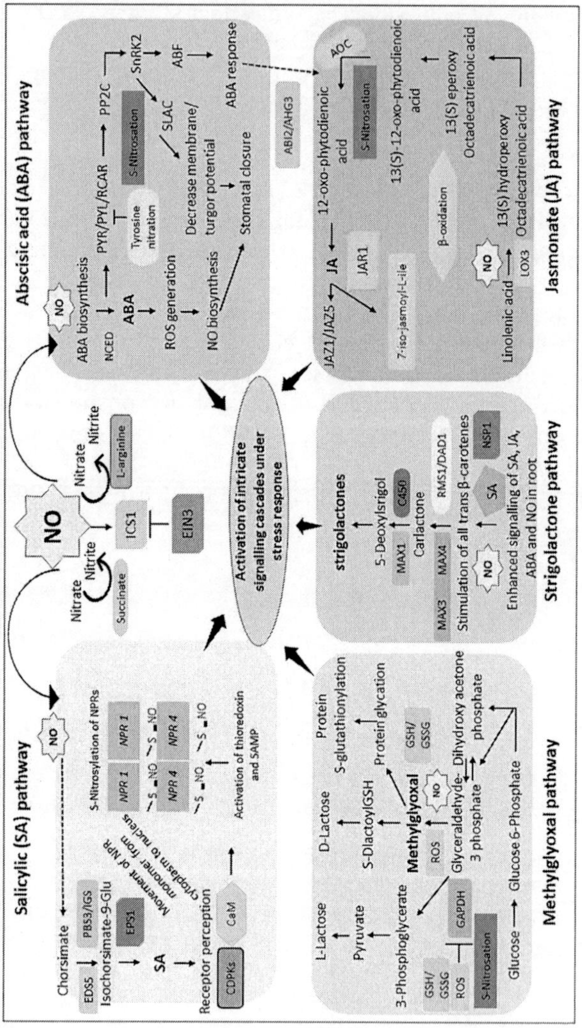


**Figure 1:** An interactive representation of salicylic acid, nitric oxide, abscisic acid, jasmonic acid, methylglyoxal and strigolactones signaling cascades. Summary versions of biosynthetic pathways of salicylic acid, nitric oxide, abscisic acid, jasmonic acid, methylglyoxal and strigolactones and mechanisms of their coordinative functions in modulating defence related genes/enzymes to improve growth and immunity of plants under stress conditions. EDS5: Enhanced disease susceptibility 5, PBS3/IGS: AVRPPHB susceptible 3 /isochorismoyl-glutamate synthase, EPS1: Enhanced pseudomonas susceptibility 1, CDPKs: Calcium dependent

protein kinases, CaM: Calmodulin, S-NO: S-nitrosation, NPR1: Non-expresser of pathogenesis related protein 1, ICS1: Isochorsimate synthase 1, EIN3: Ethylene insensitive 3, NCED: Nine-cis-epoxycarotenoid dioxygenase, SLAC: Slow anion channel, PP2C: Protein phosphatase 2C, SnRK2: Sucrose nonfermenting 1 (SNF1)-related protein kinase 2, ABF: Abscisic acid responsive factor, LOX3: Lipoxygenase 3, JAR1: Jasmonate resistance 1, AOC: Allene oxide cyclase, ABI2: Abscisic acid insensitive 1, AHG3: Member of PP2C protein family, JAZ1 and 5: Jasmonate ZIM-Domain 1 and 5 , NSP1: Nodulation signaling pathway 1, MAX 1,3,4: More auxiliary growth 1, 3, 4, C450: Cytochrome P450, RMS1/DAD1: Carotenoid cleavage dioxygenase family gene ROS: Reactive oxygen species, GSH/GSSG: Reduced glutathione/oxidized glutathione (Adapted from Rai et al., 2021)

## Future Perspectives and Challenges

The study of the role of SA and NO in stress responses has provided valuable insights into plant physiology and stress adaptation. However, several future perspectives and challenges still need to be addressed. Firstly, further research is required to unravel the intricate molecular mechanisms underlying the cross-talk between SA and NO signaling pathways. Although significant progress has been made in understanding their functions, the precise interactions and downstream signaling events remain elusive. Investigating the key components and regulatory networks involved in this cross-talk will provide a deeper understanding of how SA and NO cooperate in stress responses (Rai et al., 2021; Delaney et al., 1994).

Additionally, exploring the cross-regulation between SA and NO with other signaling molecules and pathways will broaden our understanding of the complex signaling networks involved in stress responses. Integrating SA-NO signaling with hormonal pathways, such as ABA, ethylene, and jasmonic acid, can provide a comprehensive picture of how plants integrate multiple signals to cope with stress (Kazan, 2015; Wasternack and Song, 2017). Moreover, investigating the interplay between SA-NO signaling and other signaling pathways, such as ROS and calcium signaling, will enhance our understanding of the interconnected nature of stress signaling (Neill et al., 2002; Steinhorst and Kudla, 2013). Furthermore, translating the knowledge gained from studying SA and NO signaling into practical applications for crop improvement is a crucial challenge. Developing strategies to enhance stress tolerance in crop plants by manipulating SA and NO signaling pathways holds great promise for agriculture. However, it requires a multidisciplinary approach, integrating molecular biology, genetics, and crop breeding techniques. Transgenic methods, genome editing techniques, and marker-assisted breeding can engineer crop plants with enhanced stress tolerance by modulating SA and NO signaling (Saruhan et al., 2012; Rai et al., 2018, 2020).

Lastly, environmental factors, such as climate change and pollutant stresses, pose additional challenges for plant stress responses. Understanding how these factors influence SA and NO signaling pathways and how they interact with other stress-responsive pathways will be essential for developing resilient crop plants capable of withstanding multiple stresses.

## Unresolved questions and gaps in our understanding of salicylic acid and nitric oxide signaling

Significant progress has been made in understanding the roles of SA and NO signaling in various biological processes; however, there are still several unresolved questions and gaps in our understanding of these signaling pathways. One of the significant gaps is the precise molecular mechanisms underlying the cross-talk between SA and NO signaling. Although it is known that SA and NO can interact and modulate each other's signaling pathways, the specific components and interactions involved are not fully elucidated. Unraveling the molecular players and the exact mechanisms of this cross-talk will provide valuable insights into integrating these signaling pathways and their coordinated responses to various stimuli (Arora et al., 2016; Wang et al., 2015).

Another unresolved question is the context-dependent nature of SA and NO signaling. Both SA and NO exhibit intricate signaling networks that can result in diverse responses depending on the specific conditions and plant species. Understanding the factors determining the outcome of SA and NO signaling, such as concentration, timing, and cellular context, is crucial for deciphering their precise roles in different physiological and stress-related processes (Vlot et al., 2009; Aghdam et al., 2022). Additionally, the functional redundancy and overlap with other signaling molecules pose challenges in interpreting the specific contributions of SA and NO in plant responses. Both SA and NO interact with multiple signaling pathways, including those mediated by hormones, reactive oxygen species, and secondary messengers. Elucidating the cross-regulation and interplay between SA-NO signaling and these pathways will provide a more comprehensive understanding of their individual and collective roles in plant biology (Gupta et al., 2011; Borrelli et al., 2021).

Furthermore, the specific downstream targets and transcriptional regulatory networks regulated by SA and NO signaling are not fully characterized. Identifying the direct targets of SA and NO and understanding their regulatory mechanisms will shed light on the specific genes and pathways modulated by these signaling molecules (Chen & Silva, 2017; Li et al., 2016). This knowledge is essential for deciphering the molecular basis of SA and NO-mediated processes and developing strategies to manipulate their activities for agricultural and biomedical applications.

## Technological advancements and emerging approaches for studying their roles

Technological advancements have contributed significantly to our understanding of SA and NO roles in various biological processes. These advancements have provided researchers with powerful tools and approaches to study the functions and mechanisms of these signaling molecules in greater detail. One significant technological advancement is the development of high-throughput omics technologies, such as genomics, transcriptomics, proteomics, and metabolomics. These techniques allow for the simultaneous analysis of thousands of genes, transcripts, proteins, and metabolites, providing a comprehensive view of the molecular changes associated with SA and NO signaling. By integrating omics data, researchers can identify vital regulatory components, signaling networks, and metabolic pathways influenced by SA and NO (Sah et al., 2016; Meena et al., 2017).

In addition, advanced imaging techniques have revolutionized our ability to visualize and localize SA and NO in plants. Fluorescent probes and biosensors specific to SA and NO have been developed, enabling real-time monitoring of their spatial and temporal distribution in plant tissues (Rai et al., 2020; Zhang et al., 2011). These imaging techniques provide valuable insights into the dynamic changes of SA and NO during different developmental stages and stress conditions. Furthermore, genome editing technologies, such as Clustered Regularly Interspaced Short Palindromic Repeats-CRISPR-associated protein 9 (Cas9) (CRISPR-Cas9), have opened up new possibilities for studying the functional roles of specific genes and regulatory elements involved in SA and NO signaling pathways. With precise genome editing, researchers can manipulate the expression of critical genes or regulatory regions to investigate their effects on SA and NO responses, providing a deeper understanding of their roles in plant biology (Li et al., 2016; Ren et al., 2017).

Emerging approaches such as systems biology and computational modeling have contributed significantly to our understanding of SA and NO signaling. Researchers can simulate and predict the behavior and dynamics of SA and NO signaling networks by integrating experimental data with mathematical models (Naika et al., 2013; Hussain and Kayani, 2020). This approach identifies vital regulatory nodes, feedback loops, and emergent properties of the signaling pathways, providing a more comprehensive understanding of their functions. Moreover, integrating multi-omics data with network analysis and machine learning approaches has enabled the discovery of novel components and interactions within SA and NO signaling networks (Zhang et al., 2011; Kumar, 2014). These computational approaches can identify hidden patterns and regulatory modules, facilitating the identification of key players and regulatory mechanisms that may have been overlooked using traditional experimental methods alone.

Technological advancements and emerging approaches have significantly enhanced our ability to study the roles of SA and NO in plant biology. These tools and techniques give researchers unprecedented insights into SA and NO signaling pathways' functions, mechanisms, and interactions. Continued technological advancements and integration of multidisciplinary approaches will further deepen our understanding of SA and NO and their contributions to plant growth, development, and stress responses.

## Opportunities for harnessing salicylic acid and nitric oxide for crop improvement and stress resilience

SA and NO are two signaling molecules that have shown great potential for harnessing crop improvement and enhancing stress resilience in plants. Understanding their roles and mechanisms of action opens up opportunities for developing strategies to improve agricultural productivity and sustainability. One possibility is utilizing SA and NO to enhance plant defense against biotic stresses, such as pathogens and pests. Both SA and NO crucially activate defense responses, including expressing defense-related genes, synthesizing antimicrobial compounds, and reinforcing physical barriers (Xue et al., 2014; Delaney et al., 1994). By manipulating the SA and NO signaling pathways, it is possible to enhance plant resistance to various pathogens and reduce yield losses caused by diseases and pests (Wang et al., 2015; Borrelli et al., 2021).

Moreover, SA and NO have been shown to contribute to abiotic stress tolerance in plants, including drought, heat, and salinity (Miura and Tada, 2014; Rai et al., 2020). These stresses are significant constraints to crop production worldwide, and developing stress-tolerant varieties is crucial for ensuring food security. Harnessing the potential of SA and NO in improving stress resilience can be achieved by understanding the molecular mechanisms underlying their roles in stress responses and developing strategies to enhance their production or signaling pathways in crops (Wang et al., 2015; Zheng et al., 2015). Additionally, the interplay between SA and NO signaling pathways offers opportunities for comprehensive stress adaptation. Both molecules can influence each other's signaling pathways, leading to synergistic effects on stress responses. Exploring the cross-regulation between SA and NO can provide insights into the molecular networks involved in stress signaling and help develop crop varieties with enhanced stress tolerance (Rai et al., 2021; Borrelli et al., 2021).

Furthermore, advancements in biotechnology and genetic engineering techniques enable crop manipulation of SA and NO pathways. Genetic engineering can enhance the production or activity of SA and NO or engineer plants with improved stress tolerance traits. This includes the overexpression of genes involved in SA or NO synthesis, the introduction of genes encoding SA- or NO-responsive transcription factors, or the modulation of SA and NO signaling components.

## Conclusion

The cross-talk between SA and NO represents an essential regulatory mechanism in plant stress responses. The interaction and interplay between these two signaling molecules have been shown to play crucial roles in plant defence against both biotic and abiotic stresses. The reciprocal regulation of SA and NO enhances the plant's ability to withstand the pressure by activating defence pathways, inducing antioxidant systems, and modulating various physiological processes. Understanding the molecular mechanisms underlying the SA-NO cross-talk provides valuable insights for developing strategies to enhance plant stress tolerance and improve crop productivity in challenging environments. Manipulating SA and NO pathways through genetic engineering or exogenous application can improve plant resilience to various stresses, including heat, drought, and pathogen attacks. However, there are still unresolved questions and gaps in our understanding of the specific molecular interactions and regulatory networks involved in the SA-NO cross-talk.

Further research is needed to elucidate the specific signaling pathways, transcriptional regulation mechanisms, and downstream target genes influenced by SA and NO. Exploring the cross-talk between SA, NO, and other signaling molecules will also contribute to a comprehensive understanding of plant stress responses. The thorough knowledge of the SA-NO cross-talk and its impact on plant stress responses opens up new opportunities for harnessing these signaling molecules in crop improvement and stress resilience. By manipulating the SA-NO pathways, it is possible to develop crops with enhanced stress tolerance, ultimately contributing to sustainable agriculture and global food security.

## References

Aghdam, M.S., Mukherjee, S., Flores, F.B., Arnao, M.B., Luo, Z. and Corpas, F.J., 2022. Functions of melatonin during postharvest of horticultural crops. Plant and Cell Physiology, 63(12), pp.1764-1786.

Ahmad, P., Abdel Latef, A.A., Hashem, A., Abd_Allah, E.F., Gucel, S. and Tran, L.S.P., 2016. Nitric oxide mitigates salt stress by regulating levels of osmolytes and antioxidant enzymes in chickpea. Frontiers in plant science, 7, p.347.

Arasimowicz, M. and Floryszak-Wieczorek, J., 2007. Nitric oxide as a bioactive signaling molecule in plant stress responses. Plant science, 172(5), pp.876-887.

Arasimowicz-Jelonek, M. and Floryszak-Wieczorek, J., 2014. Nitric oxide: an effective weapon of the plant or the pathogen?. Molecular Plant Pathology, 15(4), pp.406-416.

Arora, D., Jain, P., Singh, N., Kaur, H. and Bhatla, S.C., 2016. Mechanisms of nitric oxide crosstalk with reactive oxygen species scavenging enzymes during abiotic stress tolerance in plants. Free Radical Research, 50(3), pp.291-303.

Bandurska, H., 2013. Salicylic acid: an update on biosynthesis and action in plant response to water deficit and performance under drought. Salicylic Acid: Plant growth and development, pp.1-14.

Beckers, G.J.M. and Spoel, S.H., 2005. Fine-tuning plant defence signaling: salicylate versus jasmonate. Plant biology, pp.1-10.

Beligni, M.V. and Lamattina, L., 2001. Nitric oxide in plants: the history is just beginning. Plant, Cell & Environment, 24(3), pp.267-278.

Blokhina, O. and Fagerstedt, K.V., 2010. Reactive oxygen species and nitric oxide in plant mitochondria: origin and redundant regulatory systems. Physiologia plantarum, 138(4), pp.447-462.

Borrelli, V., Lanubile, A. and Marocco, A., 2021. Plant Hormones and Plant Defense Response Against Pathogens. Hormones and Plant Response, pp.1-28.

Chaki, M., Begara-Morales, J.C., Valderrama, R., Aranda-Caño, L. and Barroso, J.B., 2021. New Insights into the Functional Role of Nitric Oxide and Reactive Oxygen Species in Plant Response to Biotic and Abiotic Stress Conditions. Plant Growth and Stress Physiology, pp.215-235.

Chen, J., Wang, W.H., Wu, F.H., He, E.M., Liu, X., Shangguan, Z.P. and Zheng, H.L., 2015. Hydrogen sulfide enhances salt tolerance through nitric oxide-mediated maintenance of ion homeostasis in barley seedling roots. Scientific reports, 5(1), p.12516.

Choudhury, F.K., Rivero, R.M., Blumwald, E. and Mittler, R., 2017. Reactive oxygen species, abiotic stress and stress combination. The Plant Journal, 90(5), pp.856-867.

Christou, A., Filippou, P., Manganaris, G.A. and Fotopoulos, V., 2014. Sodium hydrosulfide induces systemic thermotolerance to strawberry plants through transcriptional regulation of heat shock proteins and aquaporin. BMC Plant Biology, 14(1), pp.1-11.

Delaney, T.P., Uknes, S., Vernooij, B., Friedrich, L., Weymann, K., Negrotto, D., Gaffney, T., Gut-Rella, M., Kessmann, H., Ward, E. and Ryals, J., 1994. A central role of salicylic acid in plant disease resistance. Science, 266(5188), pp.1247-1250.

Delangiz, N., Moghiseh, E. and Lajayer, B.A., 2022. Salicylic Acid-Mediated Regulation of Plant Biology: An Omics Approach. Managing Plant Stress Using Salicylic Acid: Physiological and Molecular Aspects, pp.16-26.

Dubiella, U., Seybold, H., Durian, G., Komander, E., Lassig, R., Witte, C.P., Schulze, W.X. and Romeis, T., 2013. Calcium-dependent protein kinase/NADPH oxidase activation circuit is required for rapid defense signal propagation. Proceedings of the National Academy of Sciences, 110(21), pp.8744-8749.

Durrant, W.E. and Dong, X., 2004. Systemic acquired resistance. Annu. Rev. Phytopathol., 42, pp.185-209.

Fancy, N.N., Bahlmann, A.K. and Loake, G.J., 2017. Nitric oxide function in plant abiotic stress. Plant, Cell & Environment, 40(4), pp.462-472.

Fu, Z.Q. and Dong, X., 2013. Systemic acquired resistance: turning local infection into global defense. Annual review of plant biology, 64, pp.839-863.

Gupta, K.J., Fernie, A.R., Kaiser, W.M. and van Dongen, J.T., 2011. On the origins of nitric oxide. Trends in plant science, 16(3), pp.160-168.

Hancock, J.T., 2020. Nitric oxide signaling in plants. Plants, 9(11), p.1550.

Hara, M., Furukawa, J., Sato, A., Mizoguchi, T. and Miura, K., 2012. Abiotic stress and role of salicylic acid in plants. Abiotic stress responses in plants: metabolism, productivity and sustainability, pp.235-251.

Hayat, S., Ali, B. and Ahmad, A., 2007. Salicylic acid: biosynthesis, metabolism and physiological role in plants. Salicylic acid: A plant hormone, pp.1-14.

Herrera-Vásquez, A., Salinas, P. and Holuigue, L., 2015. Salicylic acid and reactive oxygen species interplay in the transcriptional control of defense genes expression. Frontiers in plant science, 6, p.171.

Horváth, E., Szalai, G. and Janda, T., 2007. Induction of abiotic stress tolerance by salicylic acid signaling. Journal of Plant Growth Regulation, 26, pp.290-300.

Hou, S. and Tsuda, K., 2022. Salicylic acid and jasmonic acid crosstalk in plant immunity. Essays in Biochemistry, 66(5), pp.647-656.

Hussain, F. and Kayani, H.U.R., 2020. Aging-Oxidative stress, antioxidants and computational modeling. Heliyon, 6(5).

Janda, T., Szalai, G. and Pál, M., 2020. Salicylic acid signaling in plants. International Journal of Molecular Sciences, 21(7), p.2655.

Jiang, C., Iu, B. and Singh, J., 1996. Requirement of a CCGAC cis-acting element for cold induction of the BN115 gene from winter Brassica napus. Plant molecular biology, 30, pp.679-684.

Joudoi, T., Shichiri, Y., Kamizono, N., Akaike, T., Sawa, T., Yoshitake, J., Yamada, N. and Iwai, S., 2013. Nitrated cyclic GMP modulates guard cell signaling in Arabidopsis. The Plant Cell, 25(2), pp.558-571.

Kang, G., Li, G. and Guo, T., 2014. Molecular mechanism of salicylic acid-induced abiotic stress tolerance in higher plants. Acta Physiologiae Plantarum, 36, pp.2287-2297.

Kazan, K., 2015. Diverse roles of jasmonates and ethylene in abiotic stress tolerance. Trends in plant science, 20(4), pp.219-229.

Khan, M.N., Mobin, M., Mohammad, F. and Corpas, F.J. eds., 2015. Nitric oxide action in abiotic stress responses in plants (No. 11446). Basel, Switzerland:: Springer International Publishing.

Kolbert, Z., Feigl, G., Freschi, L. and Poór, P., 2019. Gasotransmitters in action: nitric oxide-ethylene crosstalk during plant growth and abiotic stress responses. Antioxidants, 8(6), p.167.

Kumar, D., 2014. Salicylic acid signaling in disease resistance. Plant Science, 228, pp.127-134.

Kumar, D., Yusuf, M.A., Singh, P., Sardar, M. and Sarin, N.B., 2013. Modulation of antioxidant machinery in α-tocopherol-enriched transgenic Brassica juncea plants tolerant to abiotic stress conditions. Protoplasma, 250, pp.1079-1089.

Li, J., Meng, X., Zong, Y., Chen, K., Zhang, H., Liu, J., Li, J. and Gao, C., 2016. Gene replacements and insertions in rice by intron targeting using CRISPR–Cas9. Nature plants, 2(10), pp.1-6.

Li, N., Han, X., Feng, D., Yuan, D. and Huang, L.J., 2019. Signaling crosstalk between salicylic acid and ethylene/jasmonate in plant defense: do we understand what they are whispering?. International journal of molecular sciences, 20(3), p.671.

Li, N., Han, X., Feng, D., Yuan, D. and Huang, L.J., 2019. Signaling crosstalk between salicylic acid and ethylene/jasmonate in plant defense: do we understand what they are whispering?. International journal of molecular sciences, 20(3), p.671.

Li, Y., Wu, Y., Liao, W., Hu, L., Dawuda, M.M., Jin, X., Tang, Z., Yang, J. and Yu, J., 2020. Nitric oxide is involved in the brassinolide-induced adventitious root development in cucumber. BMC plant biology, 20(1), pp.1-12.

Mata, C.G. and Lamattina, L., 2001. Nitric oxide induces stomatal closure and enhances the adaptive plant responses against drought stress. Plant Physiology, 126(3), p.1196.

Meena, K.K., Sorty, A.M., Bitla, U.M., Choudhary, K., Gupta, P., Pareek, A., Singh, D.P., Prabha, R., Sahu, P.K., Gupta, V.K. and Singh, H.B., 2017. Abiotic stress responses and microbe-mediated mitigation in plants: the omics strategies. Frontiers in plant science, 8, p.172.

Mittler, R., Finka, A. and Goloubinoff, P., 2012. How do plants feel the heat?. Trends in biochemical sciences, 37(3), pp.118-125.

Miura, K. and Tada, Y., 2014. Regulation of water, salinity, and cold stress responses by salicylic acid. Frontiers in plant science, 5, p.4.

Mur, L.A., Mandon, J., Persijn, S., Cristescu, S.M., Moshkov, I.E., Novikova, G.V., Hall, M.A., Harren, F.J., Hebelstrup, K.H. and Gupta, K.J., 2013. Nitric oxide in plants: an assessment of the current state of knowledge. AoB plants, 5, p.pls052.

Naika, M., Shameer, K., Mathew, O.K., Gowda, R. and Sowdhamini, R., 2013. STIFDB2: an updated version of plant stress-responsive transcription factor database with additional stress signals, stress-responsive transcription factor binding sites and stress-responsive genes in Arabidopsis and rice. Plant and Cell Physiology, 54(2), pp.e8-e8.

Nazar, R., Umar, S., Khan, N.A. and Sareer, O., 2015. Salicylic acid supplementation improves photosynthesis and growth in mustard through changes in proline accumulation and ethylene formation under drought stress. South African Journal of Botany, 98, pp.84-94.

Neill, S.J., Desikan, R., Clarke, A., Hurst, R.D. and Hancock, J.T., 2002. Hydrogen peroxide and nitric oxide as signaling molecules in plants. Journal of experimental botany, 53(372), pp.1237-1247.

Ohama, N., Sato, H., Shinozaki, K. and Yamaguchi-Shinozaki, K., 2017. Transcriptional regulatory network of plant heat stress response. Trends in plant science, 22(1), pp.53-65.

Oz, M.T., Eyidogan, F., Yucel, M. and Öktem, H.A., 2015. Functional role of nitric oxide under abiotic stress conditions. Nitric oxide action in abiotic stress responses in plants, pp.21-41.

Pardo-Hernández, M., López-Delacalle, M. and Rivero, R.M., 2020. ROS and NO regulation by melatonin under abiotic stress in plants. Antioxidants, 9(11), p.1078.

Pieterse, C.M., Van der Does, D., Zamioudis, C., Leon-Reyes, A. and Van Wees, S.C., 2012. Hormonal modulation of plant immunity. Annual review of cell and developmental biology, 28, pp.489-521.

Prakash, V., Singh, V.P., Tripathi, D.K., Sharma, S. and Corpas, F.J., 2021. Nitric oxide (NO) and salicylic acid (SA): A framework for their relationship in plant development under abiotic stress. Plant Biology, 23, pp.39-49.

Proietti, S., Bertini, L., Timperio, A.M., Zolla, L., Caporale, C. and Caruso, C., 2013. Crosstalk between salicylic acid and jasmonate in Arabidopsis investigated by an integrated proteomic and transcriptomic approach. Molecular BioSystems, 9(6), pp.1169-1187.

Rai, K.K., Pandey, N. and Rai, S.P., 2020. Salicylic acid and nitric oxide signaling in plant heat stress. Physiologia plantarum, 168(2), pp.241-255.

Rai, K.K., Pandey, N., Rai, N., Rai, S.K. and Pandey-Rai, S., 2021. Salicylic acid and nitric oxide: insight into the transcriptional regulation of their metabolism and regulatory functions in plants. Frontiers in Agronomy, 3, p.781027.

Rai, K.K., Rai, N. and Rai, S.P., 2018. Salicylic acid and nitric oxide alleviate high temperature induced oxidative damage in Lablab purpureus L plants by regulating bio-physical processes and DNA methylation. Plant Physiology and Biochemistry, 128, pp.72-88.

Ren, B., Yan, F., Kuang, Y., Li, N., Zhang, D., Lin, H. and Zhou, H., 2017. A CRISPR/Cas9 toolkit for efficient targeted base editing to induce genetic variations in rice. Sci China Life Sci, 60(5), pp.516-519.

Rivas-San Vicente, M. and Plasencia, J., 2011. Salicylic acid beyond defence: its role in plant growth and development. Journal of experimental botany, 62(10), pp.3321-3338.

Sah, S.K., Reddy, K.R. and Li, J., 2016. Abscisic acid and abiotic stress tolerance in crop plants. Frontiers in plant science, 7, p.571.

Saruhan, N., Saglam, A. and Kadioglu, A., 2012. Salicylic acid pretreatment induces drought tolerance and delays leaf rolling by inducing antioxidant systems in maize genotypes. Acta Physiologiae Plantarum, 34, pp.97-106.

Singh, R.K., Prasad, A., Maurya, J. and Prasad, M., 2022. Regulation of small RNA-mediated high temperature stress responses in crop plants. Plant Cell Reports, pp.1-9.

Singhal, R.K., Jatav, H.S., Aftab, T., Pandey, S., Mishra, U.N., Chauhan, J., Chand, S., Indu, Saha, D., Dadarwal, B.K. and Chandra, K., 2021. Roles of nitric oxide in conferring multiple abiotic stress tolerance in plants and crosstalk with other plant growth regulators. Journal of Plant Growth Regulation, pp.1-26.

Steinhorst, L. and Kudla, J., 2013. Calcium and reactive oxygen species rule the waves of signaling. Plant physiology, 163(2), pp.471-485.

Van der Does, D., Leon-Reyes, A., Koornneef, A., Van Verk, M.C., Rodenburg, N., Pauwels, L., Goossens, A., Körbes, A.P., Memelink, J., Ritsema, T. and Van Wees, S.C., 2013. Salicylic acid suppresses jasmonic acid signaling downstream of SCFCOI1-JAZ by targeting GCC promoter motifs via transcription factor ORA59. The Plant Cell, 25(2), pp.744-761.

Verma, N., Tiwari, S., Singh, V.P. and Prasad, S.M., 2020. Nitric oxide in plants: an ancient molecule with new tasks. Plant Growth Regulation, 90, pp.1-13.

Vlot, A.C., Dempsey, D.M.A. and Klessig, D.F., 2009. Salicylic acid, a multifaceted hormone to combat disease. Annual review of phytopathology, 47, pp.177-206.

Wahid, A., Gelani, S., Ashraf, M. and Foolad, M.R., 2007. Heat tolerance in plants: an overview. Environmental and experimental botany, 61(3), pp.199-223.

Wang, P., Du, Y., Hou, Y.J., Zhao, Y., Hsu, C.C., Yuan, F., Zhu, X., Tao, W.A., Song, C.P. and Zhu, J.K., 2015. Nitric oxide negatively regulates abscisic acid signaling in guard cells by S-nitrosylation of OST1. Proceedings of the National Academy of Sciences, 112(2), pp.613-618.

Wasternack, C. and Song, S., 2017. Jasmonates: biosynthesis, metabolism, and signaling by proteins activating and repressing transcription. Journal of Experimental Botany, 68(6), pp.1303-1321.

Wendehenne, D., Durner, J. and Klessig, D.F., 2004. Nitric oxide: a new player in plant signaling and defence responses. Current opinion in plant biology, 7(4), pp.449-455.

Xue, R.F., Wu, J., Wang, L.F., Blair, M.W., Wang, X.M., De Ge, W., Zhu, Z.D. and Wang, S.M., 2014. Salicylic acid enhances resistance to Fusarium oxysporum f. sp. phaseoli in common beans (Phaseolus vulgaris L.). Journal of Plant Growth Regulation, 33, pp.470-476.

Yan, S. and Dong, X., 2014. Perception of the plant immune signal salicylic acid. Current opinion in plant biology, 20, pp.64-68.

Zhang, A., Zhang, J., Zhang, J., Ye, N., Zhang, H., Tan, M. and Jiang, M., 2011. Nitric oxide mediates brassinosteroid-induced ABA biosynthesis involved in oxidative stress tolerance in maize leaves. Plant and Cell Physiology, 52(1), pp.181-192.

Zhang, Y., Wang, L., Liu, Y., Zhang, Q., Wei, Q. and Zhang, W., 2006. Nitric oxide enhances salt tolerance in maize seedlings through increasing activities of proton-pump and Na+/H+ antiport in the tonoplast. Planta, 224, pp.545-555.

Zheng, L.P., Zhang, B., Zou, T., Chen, Z.H. and Wang, J.W., 2010. Nitric oxide interacts with reactive oxygen species to regulate oligosaccharide-induced artemisinin biosynthesis in Artemisia annua hairy roots. J Med Plants Res, 4(9), pp.758-766.

Zheng, X.Y., Zhou, M., Yoo, H., Pruneda-Paz, J.L., Spivey, N.W., Kay, S.A. and Dong, X., 2015. Spatial and temporal regulation of biosynthesis of the plant immune signal salicylic acid. Proceedings of the National Academy of Sciences, 112(30), pp.9166-9173.

Zhou, X., Joshi, S., Khare, T., Patil, S., Shang, J. and Kumar, V., 2021. Nitric oxide, crosstalk with stress regulators and plant abiotic stress tolerance. Plant Cell Reports, 40(8), pp.1395-1414.

Zhu, H., Ai, H., Hu, Z., Du, D., Sun, J., Chen, K. and Chen, L., 2020. Comparative transcriptome combined with metabolome analyses revealed key factors involved in nitric oxide (NO)-regulated cadmium stress adaptation in tall fescue. BMC genomics, 21, pp.1-13.